A Holter for Parkinson's Disease Motor Symptoms: STAT-ON™

RIVER PUBLISHERS SERIES IN BIOMEDICAL ENGINEERING

Series Editor

DINESH KANT KUMAR
RMIT University,
Australia

The "River Publishers Series in Biomedical Engineering" is a series of comprehensive academic and professional books which focus on the engineering and mathematics in medicine and biology. The series presents innovative experimental science and technological development in the biomedical field as well as clinical application of new developments.

Books published in the series include research monographs, edited volumes, handbooks and textbooks. The books provide professionals, researchers, educators, and advanced students in the field with an invaluable insight into the latest research and developments.

Topics covered in the series include, but are by no means restricted to the following:

- Biomedical engineering
- Biomedical physics and applied biophysics
- Bio-informatics
- Bio-metrics
- Bio-signals
- Medical Imaging

For a list of other books in this series, visit www.riverpublishers.com

A Holter for Parkinson's Disease Motor Symptoms: STAT-ON™

Editors

Joan Cabestany

Universitat Politècnica de Catalunya UPC,
Barcelona, Spain

Angels Bayés

Unitat Parkinson, Centro Médico Teknon,
Barcelona, Spain

Published 2024 by River Publishers
River Publishers
Alsbjergvej 10, 9260 Gistrup, Denmark
www.riverpublishers.com

Distributed exclusively by Routledge
605 Third Avenue, New York, NY 10017, USA
4 Park Square, Milton Park, Abingdon, Oxon OX14 4RN

A Holter for Parkinson's Disease Motor Symptoms: STAT-ON™ / Joan Cabestany and Angels Bayés.

©2024 River Publishers. All rights reserved. No part of this publication may be reproduced, stored in a retrieval systems, or transmitted in any form or by any means, mechanical, photocopying, recording or otherwise, without prior written permission of the publishers.

Routledge is an imprint of the Taylor & Francis Group, an informa business

ISBN 978-87-7004-013-6 (hardback)
ISBN 978-10-0381-227-2 (online)
ISBN 978-1-032-63286-5 (ebook master)

While every effort is made to provide dependable information, the publisher, authors, and editors cannot be held responsible for any errors or omissions.

©The Editor(s) (if applicable) and The Author(s) 2024. This book is published open access.

Open Access

This book is distributed under the terms of the Creative Commons Attribution-Non-Commercial 4.0 International License, CC-BY-NC 4.0) (http://creativecommons.org/licenses/by/4.0/), which permits use, duplication, adaptation, distribution and reproduc-tion in any medium or format, as long as you give appropriate credit to the original author(s) and the source, a link is provided to the Creative Commons license and any changes made are indicated. The images or other third party material in this book are included in the work's Creative Commons license, unless indicated otherwise in the credit line; if such material is not included in the work's Creative Commons license and the respective action is not permitted by statutory regulation, users will need to obtain permission from the license holder to duplicate, adapt, or reproduce the material.

The use of general descriptive names, registered names, trademarks, service marks, etc. in this publication does not imply, even in the absence of a specific statement, that such names are exempt from the relevant protective laws and regulations and therefore free for general use.

The publisher, the authors and the editors are safe to assume that the advice and infor-mation in this book are believed to be true and accurate at the date of publication. Neither the publisher nor the authors or the editors give a warranty, express or implied, with respect to the material contained herein or for any errors or omissions that may have been made.

Contents

Preface — xi

List of Contributors — xiii

List of Figures — xv

List of Tables — xxi

List of Abbreviations — xxiii

1 **Management of Parkinson's Disease: Challenges and Solutions** — 1
   Angels Bayés and Jorge Hernandez-Vara
   1.1 Introduction — 2
   1.2 Strategies to Manage Parkinson's Disease at Different Stages — 4
       1.2.1 Patients at early stages — 5
       1.2.2 Moderately affected patients — 6
       1.2.3 Severely affected patients — 8
   1.3 Impact on the Quality of Life — 8
   1.4 State of the Art of the Current Trends in Illness Management — 10
   1.5 Challenges for the Best PD Management — 12
   1.6 Conclusion — 16
   References — 17

2 **Summary of the REMPARK Project Findings: Innovative Steps** — 21
   Joan Cabestany
   2.1 Summary of the REMPARK Project: Objectives, Development, and Findings — 21

2.2 Innovative Technology: Analysis of the Opportunity and
Related Challenges 24
    2.2.1 Analysis of the opportunity 27
    2.2.2 Related challenges. 31
2.3 The PARK-IT Project: Main Conclusions. 32
    2.3.1 The following steps 33
2.4 Conclusion. 34
References. 34

3 The STAT-ON™ Industrialization Pathway: From the Research Prototype to the Product   37

Carlos Pérez López, Daniel Rodríguez-Martín, and Martí Pie

3.1 Introduction 37
3.2 The Requirements of the STAT-ON™ System 39
3.3 The STAT-ON™ Hardware Electronics. 42
    3.3.1 The main circuit 42
    3.3.2 The power system 45
    3.3.3 The inertial sensors 51
    3.3.4 The printed circuit board (PCB). 51
3.4 The STAT-ON™ Firmware 53
    3.4.1 Firmware for the Nordic nRF51822. 53
    3.4.2 Firmware for the STM32F415RGT6 62
3.5 Device Mechanical Design 63
    3.5.1 Components selection 64
    3.5.2 Enclosure industrialization 68
    3.5.3 The belt 69
    3.5.4 Packaging and labeling 72
    3.5.5 Battery charging system 73
3.6 Certification and Characteristics 74
3.7 Conclusions 76
References. 80

4 The EU Medical Device Regulatory Process: The STAT-ON™   81

Daniel Rodríguez-Martín and Martí Pie

4.1 Introduction 81
    4.1.1 Definition of a medical device. 81
    4.1.2 Directive MDD93/42 and the regulation
          MDR2017/745. 82
    4.1.3 The regulation processes 84

	4.2	The Manufacturer's License.	86
	4.3	The Technical Documentation.	88
		4.3.1 Part A: The summary	88
		4.3.2 Part B: The detailed documentation.	90
		4.3.3 Part C: Updates/device modifications	103
	4.4	Quality Management System	104
	4.5	Conclusions	115
	References		116
5	**STAT-ON™: The User Interface and Generated Report**		**117**

Daniel Rodríguez-Martín, Carlos Pérez-López, Martí Pie, and Albert Pagès

	5.1	Introduction	117
	5.2	Requirements, Interface Description, and Different Modes of the Device.	118
		5.2.1 The physical interface (HMI)	120
		5.2.2 The sensor modes	122
		5.2.3 The software interface (HCI)	124
	5.3	The Application (App) and Its Management	125
		5.3.1 HOME: Main screen	126
		5.3.2 The reports.	131
	5.4	Report Hints and Interpretation	145
		5.4.1 Some interpretations on the weekly summary of motor state graph	145
		5.4.2 Some details on the weekly FoG episodes	148
		5.4.3 Some recommendations for a correct use of STAT-ON™	148
	5.5	Conclusion.	149
	References		149
6	**STAT-ON™: The Holter for Parkinson's Disease Motor Symptoms. Real Use Cases in Clinical Praxis**		**151**

Núria Caballol, Angels Bayés, Anna Planas-Ballvé, Tània Delgado, Asunción Avila, Alexandra Pérez-Soriano, López-Ariztegui Núria, Sònia Escalante, Diego Santos-Garcia, Jaime Herreros, Iria Cabo, Jorge Hernandez-Vara, José Maria Barrios, Lucía Triguero, Alvaro García-Bustillo, and Esther Cubo

	6.1	Introduction	152
	6.2	Early Detection of Motor Fluctuations	153

6.3	Improving Awareness of the First Motor Fluctuations	155
6.4	Complimenting a Poor Patient's Interview about Her Motor Complications	158
6.5	Indirect Detection of Probable PD Nonmotor Fluctuations (NMF)	161
6.6	Deciphering the Patient's Complaints using STAT-ON™	165
6.7	Ambulatory Monitorization of a Patient with Advanced PD	171
6.8	Improvement of the Patient's Awareness of the Advanced PD Stage and the Need for a Second-line Treatment	176
6.9	Identification of CANDIDATES to a Device-aided Therapy	179
6.10	STAT-ON™ Use for LCIG Tube Adjustment	184
6.11	Monitoring FoG and Second-line Treatment	187
6.12	Improving Motor Fluctuations with Variable Flow of Apomorphine Subcutaneous Infusion: The Role of STAT-ON™	190
6.13	Simultaneous Recording of Motor Activity with the STAT-ON™ Device and Subthalamic Nucleus Field Potentials (Percept™) in Parkinson's Disease	194
6.14	Telemedicine in Parkinson's Disease: The Role of STAT-ON™	198
6.15	Conclusion	201
	References	201

7 New Open Scenarios for STAT-ON™: The Medical Perspective   **207**

Núria Caballol and Diego Santos-Garcia

7.1	Introduction	207
7.2	Detection of the First PD Motor Fluctuations	209
7.3	Identification of Freezing of Gait and Falls	210
7.4	Detection of Dyskinesias	212
7.5	Detecting Non-motor Fluctuations	213
7.6	Selection of a Patient for a Device-aided Therapy and Monitor Response	213
7.7	Monitor the Response to a Treatment	216
7.8	Use in Clinical Trials	217
7.9	Use as a Marker of Disease Progression	218

	7.10 Research and Future Scenarios with STAT-ON™.	218
	7.11 Conclusion.	219
	References.	219
8	**New Open Scenarios for STAT-ON™:**	
	The Business Perspective	**227**
	Joan Calvet and Chiara Capra	
	8.1 Introduction: A General Overview	227
	8.2 The Use of Technology for a Patient-Centered Care	229
	8.3 Market Size and Impact.	231
	8.4 STAT-ON™: The New Gold Standard	232
	8.5 Conclusions.	237
	References.	238

Index   **241**

About the Editors   **243**

Preface

Parkinson's disease (PD) is a neurodegenerative disorder with associated motor and nonmotor symptoms. It is a progressive and disabling disease with a significant impact on the quality of life.

The number of PD patients is continuously rising, adding complexity, especially in the management at the level of Public Health. It is an incurable disease with a symptomatic treatment that tries to alleviate the associated symptoms through correct medication adjustment. For this reason, it is also very important to be aware of changes in the manifestation of the symptoms, which may indicate the need for an adjustment or even a change in the therapy strategy.

New ICT technologies are a growing field that can provide a solution by real-time remote monitoring of the patients, giving additional objective information to the neurologists. In this way, new possibilities are opened for a more effective treatment, more accurate control in clinical trials, and early detection of motor complications.

In this domain, the present book explains the following experience from the achieved results in the REMPARK project research (presented in the River Publishers book '*Parkinson´s Disease Management through ICT: The REMPARK Approach*' ISBN 978-87-93519-46-6 (hardback) /978-87-93519-45-9 (eBook)) till the consecution of a new medical product launched to the European market (STAT-ON™) through the execution of the EU-SME Instruments Phase II project PARK-IT (contract 756861) owned by the Sense4Care S.L. company.

The new medical device STAT-ON™ is a real Holter for the motor symptoms associated with PD. It provides objective information about the severity and distribution of PD motor symptoms and their fluctuations in daily life, allowing for unbiased and correct monitoring of the patient.

The book covers the following aspects:

- Discussion about new PD management style using the appropriate technology.

- Several clinical experiences using STAT-ON™ are reported by neurologists and movement disorders experts from different Spanish hospitals.
- Presentation of the current European regulatory scenario for medical devices and the specific case of STAT-ON™.
- Description of the followed industrialization process in order to obtain the new commercial product.
- A description of the included user interface.
- Some concluding remarks on new clinical and related business scenarios.

Intensive complementary use of STAT-ON™ by neurologists, health professionals, and researchers will increase the independence and quality of life of patients, improving their disease management and contributing to a deeper understanding of the nature of the disease.

List of Contributors

Avila, Asunción, *Departament de Neurologia, Complex Hospitalari Moisès Broggi, Sant Joan Despí, Spain*

Barrios, José Maria, *Unidad de Trastornos del Movimiento, Hospital Universitario Virgen de las Nieves, Granada, Spain*

Bayés, Angels, *Unitat de Parkinson i Transtorns de Moviment, Centro Médico TEKNON, Barcelona, Spain*

Caballol, Núria, *Departament de Neurologia, Complex Hospitalari Moisès Broggi, Sant Joan Despí, Spain; Unitat de Parkinson i Transtorns de Moviment, Centro Médico TEKNON, Barcelona, Spain*

Cabestany, Joan, *Universitat Politècnica de Catalunya, Barcelona, Spain*

Cabo, Iria, *Complexo Hospitalario Universitario de Pontevedra, Spain*

Calvet, Joan, *Sense4Care SL, Cornellà de Llobregat, Spain*

Capra, Chiara, *Sense4Care SL, Cornellà de Llobregat, Spain*

Cubo, Esther, *HUBU – Hospital Universitario de Burgos, Spain*

Delgado, Tània, *Hospital Parc Taulí, Sabadell, Spain*

Escalante, Sònia, *Hospital Verge de la Cinta, Tortosa, Spain*

García-Bustillo, Alvaro, *HUBU – Hospital Universitario de Burgos, Spain*

Hernandez-Vara, Jorge, *Departamento de Neurología, Grupo de Investigación Enfermedades Neurodegenerativas, Campus Universitario Vall d'Hebrón, Barcelona, Spain*

Herreros, Jaime, *Hospital Universitario Infanta Leonor, Madrid, Spain*

López-Ariztegui, Núria, *Unidad de Trastornos del Movimiento, Hospital Universitario de Toledo, Spain*

Pagès, Albert, *Sense4Care S.L, Cornellà de Llobregat, Spain*

Pérez-López, Carlos, *Sense4Care S.L, Cornellà de Llobregat, Spain; Consorci Sanitari de l'Alt Penedès i Garraf, Research Department, Spain*

Pérez-Soriano, Alexandra, *Unitat de Parkinson i Transtorns de Moviment, Centro Médico TEKNON, Barcelona, Spain; Fundació de Recerca Clinic Barcelona, Hospital Clinic de Barcelona, Barcelona, Spain*

Pie, Martí, *Sense4Care S.L, Cornellà de Llobregat, Spain*

Planas-Ballvé, Anna, *Departament de Neurologia, Complex Hospitalari Moisès Broggi, Sant Joan Despí, Spain*

Rodríguez-Martín, Daniel, *Sense4Care S.L, Cornellà de Llobregat, Spain*

Santos-Garcia, Diego, *CHUAC – Complejo Hospitalario Universitario de A Coruña, A Coruña, Spain; Departamento de Neurología, Hospital San Rafael, A Coruña, Spain*

Triguero, Lucía, *Unidad de Trastornos del Movimiento, Hospital Universitario Virgen de las Nieves, Granada, Spain*

List of Figures

Figure 1.1	Treatment algorithm in early Parkinson's disease patients.	4
Figure 1.2	Decision tree algorithm to manage advanced Parkinson's disease..	11
Figure 2.1	Redesigned PARK-IT wearable prototype.	34
Figure 3.1	STAT-ON™ data flow scheme (patient is wearing the sensor for a period of 7 days while doing normal activities).	40
Figure 3.2	Sensor device architecture.	43
Figure 3.3	The power system and regulator management scheme.	46
Figure 3.4	WPC V1.2 receiver power system.	48
Figure 3.5	JEITA guidelines for charging Li-ion batteries (notebook applications).	49
Figure 3.6	Circuit to control the battery's temperature..	50
Figure 3.7	Antenna connection and isolation using vias connected to ground.	52
Figure 3.8	Crystal circuit rings.	53
Figure 3.9	3D circuit model views..	53
Figure 3.10	nRF51822 system with related modules.	54
Figure 3.11	nRF firmware structure FSM..	55
Figure 3.12	ST firmware code architecture.	62
Figure 3.13	View of resistive keypad.	65
Figure 3.14	Sealing strip shape and dimensions.	66
Figure 3.15	Insert performance.	67
Figure 3.16	Designed housing. Initial design.	68
Figure 3.17	Housing overall dimensions.	69
Figure 3.18	View of the different parts forming the complete housing.	70
Figure 3.19	General measurements of the box.	70
Figure 3.20	Bottom view of the enclosure showing the part of the screws..	71

List of Figures

Figure 3.21	Ejection points.	71
Figure 3.22	Final industrialized enclosure of the STAT-ON™ device.	72
Figure 3.23	The belt.	73
Figure 3.24	The STAT-ON™ packaging.	74
Figure 3.25	Assembled packaging ready to go.	75
Figure 3.26	The STAT-ON™ labeling.	75
Figure 3.27	Charging pad aspect and dimensions.	76
Figure 4.1	Main processes and steps for achieving the EC certificate.	84
Figure 4.2	Whole detailed diagram process for achieving the EC Certificate.	85
Figure 4.3	Manufacturer's license documentation.	86
Figure 4.4	Structure of the technical documentation.	89
Figure 4.5	The schematics example corresponds to the nRF51822 processor.	93
Figure 4.6	STAT-ON™ labeling	94
Figure 4.7	Sensor's interface.	95
Figure 4.8	Weekly motor state report. The button pressed can indicate an intake of the medication.	96
Figure 4.9	PDCA approach to the QMS.	105
Figure 4.10	Involved processes in quality system.	106
Figure 4.11	Company's structure as described in the quality manual.	110
Figure 5.1	Physical interface.	120
Figure 5.2	Device correctly placed on the charging platform (LED always orange while charging and always off when the charge is completed).	123
Figure 5.3	Software architecture.	124
Figure 5.4	Main screen aspect.	127
Figure 5.5	Configuration menu.	128
Figure 5.6	Synchronization menu,	130
Figure 5.7	Summary page example.	134
Figure 5.8	Weekly motor state. The presence of FoG is indicated. In this case, the button was pressed at medication intake.	136
Figure 5.9	Time in OFF state.	137
Figure 5.10	Weekly FoG episode detection.	138
Figure 5.11	Weekly bradykinesia index (stride fluidity).	139
Figure 5.12	Clinical evaluation guideline.	140

List of Figures xvii

Figure 5.13	Weekly average cadence.	141
Figure 5.14	Weekly average of stride speed.	142
Figure 5.15	Daily motor states, including information about FoG episodes.	143
Figure 5.16	Daily stride fluidity and motor states.	144
Figure 5.17	Daily energy expenditure and the motor states in the background.	144
Figure 5.18	Part of the weekly summary graph of a patient.	146
Figure 5.19	Example of active periods followed by inactive ones.	146
Figure 5.20	Some details related to fall detection, the existence of FoG, and pressing the button situation.	147
Figure 5.21	A false positive was observed in a healthy person wearing STAT-ON™.	147
Figure 6.1	Weekly summary reported by STAT-ON™ (it shows delayed-ON in the morning, between 7 and 8 am). Additional OFF periods appeared around 12 am, and later, between 5 and 6 pm were also detected.	155
Figure 6.2	First STAT-ON™ report. It shows OFF and intermediate periods in the morning. Intermediate periods are also shown, especially in the afternoon (however, the patient denied having motor fluctuations).	157
Figure 6.3	Second STAT-ON™ report. The patient is now aware of his morning akinesia and wearing-off episodes. An increase in OFF time and decreased FoG episodes occurred (compared to Figure 6.2).	158
Figure 6.4	Third STAT-ON™ report. In this report, it is shown that FoG is less frequent. However, in this year, the patient complains of having more intense OFF periods.	159
Figure 6.5	The STAT-ON™ report generally shows ON periods with dyskinesias in the morning, followed by OFF period between 1 and 5 pm. Around 6 pm, she experiences ON periods with dyskinesias again.	161
Figure 6.6	STAT-ON™ report showing a predominant OFF period between 2 and 7 pm, in coincidence with the experienced nonmotor fluctuation.	163

xviii   List of Figures

Figure 6.7	Percentage and number of OFF hours per day detected by STAT-ON™ (December 24 to 28, 2020)..	164
Figure 6.8	Percentage and number of OFF hours per day detected by STAT-ON™ 3 months later (from March 24 to 26, 2021)................	165
Figure 6.9	(A) Results of the first 3 days. After the morning akinesia period, she had ON with bothersome dyskinesia. (B) Total number of hours in the OFF time distributed per day................	168
Figure 6.10	Compared with the previous 3-day monitoring period (Figure 6.9), the report shows a better motor state with an average of 1.5 h per day in OFF state.....................	169
Figure 6.11	Patient's diary while using the sensor. In this case, the diary was useful for the patient to identify bothersome dyskinesia. The asterisks in green represented the moment the patient did not feel good, and the symptoms were bothersome to her. For example, the dyskinesia was annoying on Saturday 11, at midnight, or on Sunday 12, around 1 pm......................	170
Figure 6.12	Summary of STAT-ON™ report. As indicated, it was detected a 21% of OFF Time, a 47.3% of ON time, and a 24.6 % of intermediate state. She was suffering from dyskinesias during a 13.2% of the monitoring time. . .	174
Figure 6.13	The STAT-ON™ report shows intermediate state in the morning with FoG episodes. The report demonstrates OFF states associated with the LD/CD doses. Dyskinesias were especially present in the afternoon.....................	175
Figure 6.14	STAT-ON™ summary report showing the above-mentioned situation....................	177
Figure 6.15	Percentage of daily OFF time.............	178
Figure 6.16	STAT-ON™ report summary showing well-defined OFF time periods in the morning and after midday. Many FoG episodes were also detected........	182
Figure 6.17	Weekly FoG report showing that the worst day regarding the number of FoG episodes was on Monday 16.......................	183
Figure 6.18	Weekly summary of OFF time.............	184

Figure 6.19	LCIG tube mispositioning in stomach (left) and right positioning in ileum (right). Yellow arrow points to the tail of the internal probe.	186
Figure 6.20	Percentage of daily OFF time (A) preapomorphine and (B) postapomorphine.	190
Figure 6.21	Number of FoG episodes per day (dot indicates the average duration) (A) preapomorphine and (B) postapomorphine.	191
Figure 6.22	STAT-ON™ summary report (A) preapomorphine and (B) postapomorphine.	192
Figure 6.23	STAT-ON™ report summarizes the motor status before starting apomorphine infusion.	192
Figure 6.24	Motor status in terms of motor complications after 3 months of subcutaneous apomorphine infusion (with the constant flow).	193
Figure 6.25	Improvement of motor fluctuations, especially in the morning, after 3 months of subcutaneous apomorphine infusion with variable flow.	194
Figure 6.26	Correlation between daily recordings of beta bands and motor status according to Percept™ and STAT-ON™, respectively. Beta bands (blue spikes in the top graph) coincide with OFF (red), "intermediate" (yellow) or "not applicable" (gray = no motion detected) motor status periods detected by STAT-ON™. The ON periods (green) coincide with beta-band free intervals. Most freezing of gait (FoG) episodes detected by STAT-ON™ coincide with non-ON periods..	197

List of Tables

Table 1.1	Modified Hoehn and Yahr scale.	5
Table 2.1	Summary of the main REMPARK project results.	25
Table 2.2	Conclusions of the analysis of opportunity after REMPARK.	30
Table 3.1	Complete list of the sensor requirements.	41
Table 3.2	Battery specifications.	49
Table 3.3	LDO's dropout voltage specification.	50
Table 3.4	Electromechanical requirements.	64
Table 3.5	The IP code for STAT-ON™ is IP65.	65
Table 3.6	Inserts parameters.	67
Table 3.7	STAT-ON™ characteristics.	77
Table 3.8	Standards affecting STAT-ON™.	79
Table 4.1	Risk management list for STAT-ON™.	99
Table 5.1	Description of the main user requirements.	119
Table 6.1	Motor's state according to the patient opinion.	186
Table 6.2	Summary of the STAT-ON™ report before and after the LCIG tube mispositioning.	187
Table 6.3	Results of the assessments pre and postmultidisciplinary telemedicine program.	199
Table 6.4	Summary of the motor status measured by STAT-ON™ at the baseline visit and after 4 months of the program.	200
Table 6.5	Summary of the presented real use cases.	203

List of Abbreviations

ADL	Activities of daily living
APD	Advanced Parkinson's disease
App	Application – smartphone application.
BI	Bradykinesia Index
CAGR	Compound annual growth rate
DBS	Deep brain stimulation
DMS	Disease management system
FoG	Freezing of gait
GDPR	General data protection regulation
H and Y	Hoehn and Yahr
HCI/HMI	Human computer/machine interface
HHC	Home healthcare
I2C	Inter-integrated circuits
IFU	Instructions for use
INT State	Intermediate state
IPR	Intellectual property right
JEITA	Japan Electronics and Information Technology Industries Association
LCIG	Levodopa–carbidopa intestinal gel
LDO	Low dropout regulator
LSVT	Lee Silverman Voice Treatment
MA	Morning akinesia
MDD	Medical device directive
MDR	Medical device regulation
MDT	Multidisciplinary treatment
MF	Motor fluctuations
NMF	Non-motor fluctuation
NMS	Non-motor symptoms
PD	Parkinson's disease
PDCA	Plan-do-check-act
PMCFU	Post-market clinical follow-up
QMS	Quality management system

QoL/QoLRH	Quality of life/quality of life related to health
QoL	Quality of life
RIA	Research innovation action
SME	Small-medium enterprise
SPI	Serial peripheral interface
TCOC	Total cost of care
TRL	Technology readiness level
UART/USART	Universal Asynchronous (Serial) Receiver Transmitter
WO	Wearing-off
WOQ-19	Wearing-off19 questionnaire

1

Management of Parkinson's Disease: Challenges and Solutions

Angels Bayés[1] and Jorge Hernandez-Vara[2]

[1]UParkinson. Centro Médico Teknon, Barcelona, Spain
[2]Neurology Department and Neurodegenerative Diseases Research Group, Vall D'Hebron University Campus, Barcelona, Spain

Email: abayes@uparkinson.org; hernandezvarajorge76@gmail.com

Abstract

Parkinson's disease is the second most common neurodegenerative disorder after Alzheimer's disease. It is a progressive and disabling disease with significative impact on quality of life. Since it has no cure, available treatment is targeted to improve the symptoms due to a lack of dopamine in the central nervous system.

In this chapter, we summarized the currently available therapeutic strategies to manage the early and advanced stages of the disease.

As the disease progresses, treatment becomes more complex and it is necessary to have simple and objective tools to detect fluctuations in the motor status of patients and closely monitor their response to treatment.

Here, current difficulties and barriers to Parkinson's disease management are described. In addition, the role of new technologies is introduced as potential supporting tools to provide a more holistic approach to the treatment of the disease. For all these reasons, the need of having multidisciplinary teams accessible to the patients is also discussed.

In summary, Parkinson's disease is a complex and multisystem disorder that requires a multidisciplinary and holistic approach compressing all the aspects of the disease to improve the quality of life of the patients. New technologies are a growing field that could provide a potential solution to the current unmeet of this disabling disease by real-time remote monitoring.

1.1 Introduction

Parkinson's disease (PD) is the second most frequent neurodegenerative disorder, with approximately 6.1 million people who live with PD in 2016 worldwide [1]. For several reasons that are not yet fully understood, the prevalence and incidence are expected to increase in the next years. According to the World Health Organization, globally, disability and death due to PD are increasing faster than for any other neurological disorder [2].

There is currently no cure for PD, but there are treatments available to relieve the symptoms and maintain an individual's quality of life (QoL) at least for the first years.

The PD impact on the QoL is due to an enormous number of motor and non-motor symptoms: bradykinesia, rigidity, tremor, postural instability, reduced gait speed, freezing of gait (FoG), sleep disturbances, depression, psychosis, autonomic and gastrointestinal dysfunction as well as dementia. The majority of patients will develop an increasing number of more complex symptoms over time.

The treatment in the early stages of the disease, focused on the use of levodopa, is very effective. Nevertheless, different problems related to the treatment or disease progression may start to appear depending on the advance of the disease. Thus, it might be the case of motor complications (MCs): **motor fluctuations** such as the **wearing-off** phenomenon (temporary loss of dopaminergic effect), involuntary movements known as **dyskinesia**, fluctuations between the ON stage (when a correct control of the symptoms is achieved) and the OFF stage (when motor symptoms reappear), abnormal cramps and postures of the extremities and trunk known as **dystonia**, and a variety of complex fluctuations in other motor and nonmotor functions, the nonmotor complications (NMCs). In these cases, the precise adjustment of the therapy is crucial to avoid decreasing the QoL of the patient. The motor symptoms are especially responsible for falls and gait impairments and negatively impact on QoL by reducing the ability to perform many activities of daily living. They are the major causes of institutionalization and by the way, losing independence. Daily tasks at home (self-care, food preparation, climbing stairs…) become difficult, as do many activities outside the home such as shopping, visiting friends/family, leisure activities, among others.

The management of this disease must be multidimensional. Unfortunately, there is often no integration between data at different levels of the health system: primary health physicians, occupational therapists, and social workers. Information about the general condition of the patient is also usually lacking.

PD treatment is actually symptomatic, based on dopaminergic replacement therapy, and aims to alleviate the symptoms associated with the disease, through the precise adjustment of medication. The most widely used drug, levodopa, is effective usually across the lifespan. However, the onset of MCs, as is ON-OFF fluctuations and dyskinesias, a few years after starting treatment, detracts from its value. Symptomatic management of these complications is difficult and often needs to be optimized because the improvement obtained after this adjustment is not usually stable for a long time.

As the disease progresses, treatment is primarily addressed to reduce the time spent in the OFF state, while avoiding the appearance of MCs and NMS, such as hallucinations or impulse control disorder. Reducing OFF periods is therefore one of the main parameters used to evaluate the effectiveness of therapeutic interventions, both in medical practice and in clinical trials. Gathering accurate information about the patient's condition throughout the day is essential to determine the optimal treatment plan. In clinical practice, the only method available is based on diaries filled in by patients and their caregivers about the ON – OFF periods and dyskinesias. However, this method has certain limitations that make unreliable medium-and long-term monitoring: motor difficulties and cognitive failures that hinder regular compliance and subjective evaluation. In addition to the huge time-consuming it represents for the patient as well as the clinician, to explain how the diary should be filled out. Moreover, is one of the main reasons for screening failures in clinical trials for fluctuating PD patients. Therefore, more objective solutions that can improve disease monitoring and management are of great interest and occupy an important part of current research.

Several motor and nonmotor symptoms could appear at disease onset and over time, PD might be considered a multisystemic disease instead a pure motor disease. Thus, another important aspect of the symptomatic treatment of PD is the Multidisciplinary treatment (MDT) approach. The multiple impairments occurring in Parkinson's disease have diverse functional and psychosocial consequences. While the primary treatment is pharmacological, many symptoms do not respond well to medication, such as **ON-period freezing of gait (FoG), postural instability, speech, and swallowing difficulties**. Indeed, later-stage disease may be dominated by such symptoms. In addition, there is growing evidence for the efficacy of rehabilitation therapies and exercise for specific symptoms, through the involvement of the multidisciplinary team. There is also emerging evidence for physiotherapy with external cueing for improving gait and balance; cognitive movement strategies; and strength and balance exercises. Intensive speech therapy has been shown to improve the loudness and intelligibility of speech in Parkinson's disease.

4 Management of Parkinson's Disease

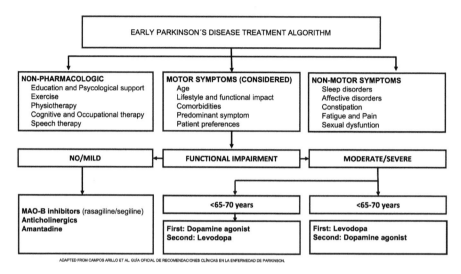

Figure 1.1 Treatment algorithm in early Parkinson's disease patients [4].

Unfortunately, the MDT is only applied in a few numbers of PD patients for economical and logistic reasons.

In the study of Winter et al. [3], a baseline and 3-, 6-, and 12-month assessments were performed on 145 Parkinson's patients. The average annual cost was calculated at 20,095 € per patient. The direct costs involved an expenditure of 13,185 € on medication, 3,526 € on hospital care, and 3,789 € on residences. The indirect costs accounted for 34.5% of the total costs (6,937 €). The costs of home care for the family accounted for 20% of direct costs. Factors associated with a higher total cost were fluctuations, dyskinesias, and younger age.

1.2 Strategies to Manage Parkinson's Disease at Different Stages

The diagnosis of PD is sometimes difficult. At the onset, the patient could show nonspecific signs, such as pain or mental depression or a slight tremor in one limb. During the first year, once the possible diagnosis is confirmed, several symptomatic treatments could be initiated depending on many factors: age, disability, and type of job, as it is summarized in Figure 1.1 [4]. Most PD patients respond very well to treatment with levodopa and dopaminergic agonists during the first years (between 3 and 7 years). This is the reason why it is called *"honeymoon period."*

Table 1.1 Modified Hoehn and Yahr scale.

Stage	Modified Hoehn and Yahr scale
1	Unilateral involvement only
1.5	Unilateral and axial involvement
2	Bilateral involvement without impairment of balance
2.5	Mild bilateral disease with recovery on pull test
3	Mild to moderate bilateral disease; some postural instability; physically independent
4	Severe disability; still able to walk or stand unassisted
5	Wheelchair-bound or bedridden unless aided

During the years 2–4, there is relative normality, and the medication is generally effective. As the disease progresses, the patient encounters a limitation of the effect of medical treatment due to the appearance of motor and non-motor complications: wearing off and dyskinesias. These entail a progressive difficulty in carrying out activities of daily living and leading an independent life. Between the years 5 through 9, the effectiveness of medication usually decreases, and treatment may need to be modified. Problems with driving, finances, and work may appear at this time. During years 10–13, there is an increasing disability: 60%–75% of patients present with some cognitive deficit, worsening immobility, incontinence, and increased risk of falls.

We can distinguish five evolutionary stages of the disease, such as the Hoehn & Yahr stages (HY) (Table 1.1). Patients do not necessarily have to go through all of them. The main problems presented by patients in the different evolutionary phases and the strategies currently recommended are considered in the following points.

1.2.1 Patients at early stages

In stage I of PD, facial expression and posture are generally normal. A tremor of a limb is the most common initial manifestation. It is often quite annoying, and it is the symptom that draws the attention of both the doctor and the patient. Typical Parkinsonian tremor appears at rest and rarely interferes with the activities of daily living (ADL), although it disturbs and distresses the patient and caregiver. Patients sometimes report difficulties in performing activities that require motor skills such as buttoning, typing, or cutting food. In the careful exploration of these patients, other Parkinsonian signs in a limb, such as bradykinesia or slow movement, and stiffness, which contribute to these fine motor difficulties, are detected in addition to tremors. Decreased arm swing or dragging of a leg when walking can also be observed. These symptoms, often present for several years, are better tolerated than tremors.

In stage II of PD, there is bilateral involvement. There may be a loss of facial expression with decreased blinking. Slight flexion of the body may be present and, in general, arm swing when walking is diminished, without altering balance. Patients slow down when performing ADLs, and they require more time to dress, clean themselves up, get up from a chair, or tie their shoes on their own.

Depressive symptoms are also frequent, and these are detected in between 30% and 50% of the cases. Medical treatment will be administered according to the severity of the symptoms. Sometimes they can produce side effects.

In these initial stages, patients are advised to learn about the disease, learn to speak naturally about their problems, learn to share difficulties and go to the doctor accompanied by someone. Standardized psychoeducational programs, such as the "Edupark" program [5] are a great help at this stage of the disease. From the diagnosis, it is recommended to initiate MDT, which includes physical exercise, and cognitive stimulation. It is better for patients to continue doing things by themselves, even if it is done slowly, without rushing, and with enough time. It is advisable to adapt the setting in which patients must perform their ADLs and to be physically and mentally as active as possible. Family members should also be informed and should know how to convey their support. It is recommended to see a doctor if depressive symptoms or side effects occur with medications.

1.2.2 Moderately affected patients

People at III-IV Hoehn and Yahr stages have already a degree of moderate-severe disability, as they experience gait and balance difficulties. They explain that their gait is shortened and that sometimes they have difficulties making turns while they walk, in the corners of the rooms, or while crossing the doors. Balance problems can cause falls. Sometimes while walking, they present FoG, or difficulty to stand, either forward (propulsion), or backward (retropulsion). The feeling of fatigue is a very frequent symptom. They have the feeling of needing a lot more effort to perform certain tasks and often notice pains in the cervical, lumbar, or shoulder region. Symptoms of autonomic dysfunction may also be present in the form of orthostatic hypotension, extreme sensations of heat or cold, sweating not related to physical activity, sometimes in the form of crisis, and urinary, gastrointestinal, or sexual dysfunction.

Many patients, at stage III or IV, experience side effects of chronic dopaminergic medication. The most annoying side effect for patients is the

ON-OFF phenomenon that can manifest with MCs or with NMCs. This phenomenon is often disabling and causes fear and insecurity. During the ON phase, patients can enjoy good mobility and carry out activities outside the home, such as shopping or social activities. However, during the OFF phase, the patient may be completely disabled, with difficulties in walking, thinking and speaking, getting up from a chair, or manipulating objects with hands among others. The appearance of OFF phases limits the social activities of the patient, often preventing them from going out with consequent worsening in terms of QoL. In this state, patients may find themselves in really dangerous situations, such as if this phenomenon occurs when crossing a street.

Dyskinesias, or abnormal involuntary movements, are another important and disabling problem that many patients present with during stages III and IV. In general, they have a choreiform nature: creeping and twisting movements of the extremities, or masticatory movements of the lower jaw, protrusion of the tongue, oscillations as they walk, and reciprocating movements with head and neck. Dyskinesias are a long-term side effect of dopaminergic medication, which usually occurs during the levodopa plasmatic peak dose. If they are mild, the family is more aware of these movements than the patients themselves, who usually associate it with the free time of Parkinsonian symptoms. When they are severe, they can become disabling as much as the symptoms themselves.

NMS may appear in form of sleep disorder, vivid dreams, and nocturnal vocalizations. Night-time vocalizations, reported by the bed partner, consist of loud cries during sleep often accompanied by the agitation of arms and legs (acting out). It is called *"REM behavior disorder."* These events can disrupt sleep. Other frequent behavioral disorders in these stages are visual hallucinations, delusional ideation, and confusing states of the paranoid type. Visual hallucinations in general are not very threatening in PD. They often describe the vision of family members, animals, or shadows that become animated objects.

The strategies recommended in these phases are aimed at understanding the MCs and NMCs and knowing how to monitor them. This will allow the patient to adjust the activities in each period. If MCs or behavioral changes appear, the neurologist can be informed to assess the possibility of adjusting the drug. It is, therefore, important to learn to do the patient's diary. This information will be crucial to optimize pharmacological treatment.

Patients at these stages should continue to maintain an active life and perform MDT, such as physical exercise, occupational therapy, speech therapy, and cognitive stimulation, according to individual needs. It is also recommended that patients continue doing things by themselves, for as long as possible.

1.2.3 Severely affected patients

Patients with PD, stage V, are severely affected. They are usually confined to a wheelchair or bedridden and require great assistance to make transfers. They are totally dependent to carry out ADLs and have a great limitation on a personal level.

Difficulties in speech and voice are often accentuated: these patients are often difficult to understand due to their low volume and poor articulation of words. They may eventually develop contractures and present decubitus ulcers or recurrent urinary tract infections.

Since the emergence of effective therapies for the treatment of this disease, not all patients reach a state of total dependence. However, they are experiencing a progressive reduction in time spent in ON and an increase in dependency time. In the lasted stages of this disease, the presence of progressive dysphagia can cause recurrent aspiration pneumonia, which is a possible cause of death. Other conditions that may contribute to this outcome are infections of pressure ulcers or urinary tract.

Since a causal treatment of the disease is still not possible, the objective for an optimal treatment will be to obtain for the patient a good QoL and the maximum independence possible. In the advanced stages, it is recommended to follow extreme hygiene, take care of mobilization, adapt the feeding, and above all take care of communication. The Lee Silverman Voice Treatment (LSVT) method has demonstrated efficacy in the treatment of speech and speech disorders. However, in very severe situations, it is advisable to maintain communication, even if external technical support is necessary.

Possible behavioral disorders should be addressed, while enhancing the hobbies and pleasures that can still take place, such as listening to music, reading, or watching movies. Caregivers should make them feel their support, while they should seek a replacement that allows them to have their own space and thus avoid the burden of care and better adaptation when the patient passes away.

1.3 Impact on the Quality of Life

PD is one of the chronic degenerative disorders with the most impact on patients' lifestyles. Most patients survive many years after the first symptoms. The mean survival rate of patients with this disease (when diagnosed after age 50) is 26 years, not very different from the nonaffected population of PD.

Quality of life (QoL) means well-being or satisfaction with aspects of life that are important to the person according to social standards and personal judgments. Because of this latter characteristic, each person understands the QoL in different ways and, therefore, it is difficult to define. The World Health Organization (1995) defines it as: *"an individual perception of the position in a person's life, in the context of the culture and value system in which he lives, in relation to his goals, expectations, standards, and concerns."*

When it is not possible to cure, maintaining the quality of life of the patient is a priority of medical care. Quality of Life, as related to Health (QoLRH), is the self-perception and assessment of the impact that the disease has on a patient's life and what its consequences are [6]. This assessment is extremely important and includes physical aspects related to the ability to perform activities, mental aspects related to mood and cognition, social aspects, and economic aspects. Several studies have been done to assess QoLRH in PD [7]. The three most important factors determining QoLRH in PD were depression, the stage of the disease, and the time that has elapsed since the onset of the disease.

In another study [8] performed with 100 patients, the most important predictor for poor QoL was depression, followed by motor complications. Motor complications, especially nocturnal akinesia, and dyskinesias, significantly decrease the QoL of Parkinson's patients [9–11]. Dyskinesias can also increase health costs in patients with PD. This should be considered when planning treatment [12].

Despite the high impact of motor symptoms in PD, nonmotor symptoms seem to influence patients' QoL even more. Nonmotor symptoms tend to accumulate. The average was 10 symptoms per patient in the populations studied and symptoms tend to intensify over time. Depression, anxiety, fatigue, sleep disorders, pain, orthostatic hypotension, and profuse sweating are some of those that have shown an individual relationship with loss of quality of life. In fact, any symptom that, due to its intensity, is installed as a central problem in the life of the patient has a direct and important impact on his/her quality of life. For example difficulty swallowing, persistent constipation, urinary urgency, night-time urination, delusions and hallucinations, memory problems, or a sense of choking when breathing. At the global level, the main factors influencing the poor QoL of those affected by PD are (in order):

1. Depression
2. Overall disease intensity (stage)

3. Dyskinesia
4. On-off fluctuations
5. Age
6. Insomnia
7. The tremor
8. Cognitive dysfunction

To assess, in a more global fashion, the impact of motor and nonmotor symptoms in terms of QoL a new staging of the disease has been proposed. The combination of HY and Nonmotor symptoms score could reflect the severity of the disease more accurately [13].

Another element that must be considered is the QoL in caregivers. Forty percent of them indicate that their health suffers due to the care. Nearly half have increased depression, and two-thirds report that their social life has suffered. The caregiver becomes burned out more (*burden of care*) if the patient has more disability, affective problems, mental confusion, or falls. There is a correlation between those caregivers that are most affected and the degree of a patient's depression and one of the main determinants of QoL in caregivers is mood changes, especially depression [14].

The conclusion is that more attention should be given to caregivers' care, particularly in advanced stages and/or with psychiatric and fall complications. These findings demonstrate that the quality of life of both the patient and the caregiver depends, to a great extent, on the inclusion of the burden of care as one of the problems associated with PD [15].

1.4 State of the Art of the Current Trends in Illness Management

As has been previously mentioned the current treatment of PD is symptomatic and is applied through pharmacological and/or surgical treatment, associated with MDT.

The pharmacological treatment of PD is focused on balancing the lack of dopamine and other neurotransmitters, and aims to alleviate the symptoms associated with the disease, by precise drug optimization. During the first years of treatment, dopaminergic drugs (levodopa and dopaminergic agonists) are usually very effective. At 2 years of levodopa treatment, 38% of patients had ON-OFF fluctuations [16].

When the ON-OFF phenomena are already present, the objective of the treatment will be essentially concentrated on reducing the time that the

1.4 State of the Art of the Current Trends in Illness Management 11

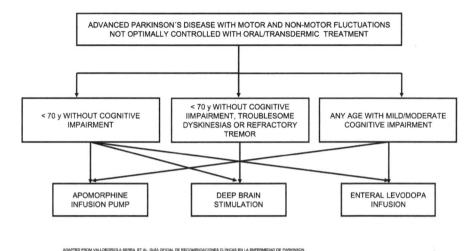

Figure 1.2 Decision tree algorithm to manage advanced Parkinson's disease.

patient spent in the OFF state. To determine the optimal and personalized treatment plan, gathering accurate information about a patient's condition throughout the day is essential. In clinical practice, the method currently available is based on clinical recall or diaries filled in by patients and their caregivers, recording hours of ON-OFF and the presence of dyskinesia. However, this method has limitations that make it unreliable in conditions of the real clinical setting, such as motor difficulties, failures in memory and in compliance, and subjective evaluation. It is necessary to know precisely and objectively the effect of drugs on the OFF stage reducing hours and increasing the ON hours in PD patients. **Reliable and easy-to-use tools are therefore needed for detecting and monitoring the motor condition of the patients.**

When both, motor and nonmotor symptoms, are not adequately controlled with oral or transdermal treatment, patients may benefit from second-line or device-aided therapies (DAT). These therapies include continuous infusions of apomorphine, enteral levodopa, and deep brain stimulation. Subcutaneous formulations of levodopa are likely to be available in the future. The main objective of these strategies is to provide continuous dopaminergic stimulation (CDS) by different mechanisms to manage and control both motor and nonmotor fluctuations. All of these therapies have shown significant efficacy in terms of increasing the quality of ON time (without troublesome dyskinesias), decreasing OFF time, and reducing the burden of nonmotor symptoms [17]. Figure 1.2 summarizes a decision tree considering the second-line therapy options [4].

However, these techniques are expensive and often difficult to manage the patient. Well-designed clinical studies on these DATs provided evidence for the efficacy of DBS and CDS in advanced PD and currently, we have new perspectives for their use also in earlier disease stages.

On the other hand, there is growing scientific evidence of the benefit of the application of MDT, such as physiotherapy, speech therapy (e.g., LSVT), occupational therapy, cognitive stimulation, and psychoeducation in the treatment of PD. Intensive and multidisciplinary rehabilitation slows the progression of motor decay and slows the need to increase treatment with levodopa, which is postulated to have a neuroprotective effect [18]. Therefore, the application of MDT from the moment of diagnosis seems of great interest. There are several studies of multidisciplinary care in Parkinson's disease comparing outcomes before and after the intervention. Outpatient multidisciplinary care programs have reported short-term improvements in motor skills, gait speed and stride length, speech, depression, and health-related quality of life. Long-term improvements in motor function have also been reported, and the authors comment that a close collaboration among members of the multidisciplinary team was essential to obtain the best results.

For the implementation of an effective multidisciplinary approach, there may be limitations, such as living far away, insufficient experience among health professionals, poor interdisciplinary collaboration, poor communication, and lack of financial support for a multidisciplinary team approach.

Regular face-to-face team meetings are important for the effective functioning of the team. These meetings allow sharing of accurate information and ensure the team is working toward shared goals for any given patient. The meetings can be a forum and stimulus for staff education, driving up the quality of care. This type of coordinated multidisciplinary approach is sometimes referred to as interdisciplinary.

Most hospitals in Europe do not have a multidisciplinary service for the care of people with PD. These types of therapies are expensive and in addition, their application requires patients to frequently go from one place to another. This entails a number of limitations, both economic and logistical, for those affected with PD before having access to these therapies.

1.5 Challenges for the Best PD Management

Current management of advanced PD is complicated and problems arising from poor quality of life affect many patients. In 2001, *the Committee on Quality of Health Care in America Institute of Medicine* provided an objective analysis of healthcare. The report listed 6 aims, proposing that health care should be: safe, effective, patient-centered, timely, efficient, and equitable. However, current care for PD in the United States, Europe, and likely the majority of the world, frequently does not meet these six aims [19]. PD care is often not safe. Individuals with PD who are hospitalized are often subjected to delayed treatment, contraindicated medications, prolonged immobility, lengthy stays, and high mortality [20, 21]. There are some comprehensive and distributed PD cares models that are quite effective, but only a few patients receive such care. Many PD-related hospitalizations are likely preventable. The patient-centered care that is timely has been rarely studied. Despite the limited evidence, focus groups and surveys suggest that individuals with PD want more personalized information from multiple disciplines that are delivered remotely in a timely manner [22]. PD care is very inefficient. Patients and their caregivers spend hours traveling and waiting in the clinic for routine follow-up appointments or for the application of complementary therapies. A recent technical publication from WHO underlines the importance of multidisciplinary teams in the holistic approach to the disease and lists a series of key actions to be implemented by health systems [2].

Finally, probably what may be most concerning, is that there exists inequity in current PD care. **A primary determinant of the care that will be received is where you live.** In the United States, 42% of individuals with PD older than 65 and up to 100% of individuals in some rural areas do not see a neurologist soon after diagnosis [23]. In Europe, the first right expressed in the European Parkinson's Disease Association Charter is care from a physician with a special interest in PD. However, 44% of Europeans do not see a PD specialist in the first 2 years after diagnosis. Beyond neurological care, access to specialist nurses, occupational therapists, and counselors is often more limited [24]. In less wealthy countries, the situation is even worse. China only has approximately 50 movement disorder specialists to care for more than 2 million individuals with PD. Bolivia only has 15 professionals. A door-to-door epidemiology study found that none of the individuals identified with PD had ever seen a physician, much less received treatment.

It is possible to make the treatment safer, effective, patient-centered, efficient, and equitable only if the following two conditions apply [25]:

- The treatment is applied, mostly, at the patient's home.
- Tools, based on new technologies, are used (sensors, communication platforms, and smartphones).

These conditions will overcome economic barriers and physical distance. However, the **potential digital gap** and the **population's access to such**

technological resources should be considered when developing health program policies.

The simple fact of detecting accurately and reliably the clinical condition of the patient can mean a 360°-change in the QoL of the patient, as this will affect a much more accurate adjustment of medication. In addition, with the help of adequate platforms, many more patients, as well as their caregivers, will receive more specialized medical care, complementary therapies, and psychoeducation as often as necessary, regardless of where they live.

In addition, **reliable detection of the motor status of PD patients throughout the day can drastically change the value of drug clinical trials**. Finally, the careful selection of patients amenable to semi-invasive therapy options becomes more and more important and should be timely. An interdisciplinary setting is required to account for optimal patient information and awareness, selection of best individual treatment modality, training of relatives and caregivers, management of complications, and follow-up care.

From a clinical point of view, the development of new technologies in the management of Parkinson's disease must be validated so that the improvement of the QoL related to health is the main objective. Symptoms' monitoring tools should be based on the following premises [26]:

- They provide a valid and accurate parameter of a clinically relevant characteristic of the disease.
- The evidence that the parameter has an ecologically relevant effect on the specific clinical application is found.
- A target interval can be defined in which the parameter reflects the appropriate treatment response.
- The implementation is simple to allow repetitive use.

Remote monitoring from devices, such as wearable sensors, smartphones, platforms, disease management applications, smart beds, wall-mounted cameras, smart glasses, etc., can monitor a patient's symptoms and function objectively in their environment, facilitating the delivery of highly personalized care.

Another aspect to improve PD care is that most of it must be done at home. Current care models frequently require caring for older people with mobility and cognition problems, with loss of the ability to drive, and needing to be transferred by overburdened caregivers, to large and complex urban medical centers.

Moving care to the patient's home would make PD care more patient-centered. Demographic factors, including aging populations, and social factors, such as the splintering of the extended family, will increase the need for home-based care.

Technological advances, especially the ability to assess and deliver care remotely, will enable the transition of care back to the home. However, despite its promise, this next generation of home-based care will have to overcome barriers, including outdated insurance models and a technological divide. Once these barriers are addressed, home-based care will increase access to high-quality care for the growing number of individuals with PD.

Emerging care models will combine remote monitoring, self-monitoring, and multidisciplinary care to enable the provision of patient-centered care at home and decrease the need for in-clinic assessments. The demand for in-home care is likely to grow as a result of demographic, economic, social, and technological factors. Both the absolute number and proportion of older individuals with PD will increase.

A system for PD management will be necessary in the near future. It must be able to reliably assess the symptoms, facilitate patient disease management, and give them independence and the best QoL. At the same time, the tools must help the patient to stay physically and mentally active as much as possible. Finally, they must provide the neurologist with disease management tools to optimize the treatment.

Emerging available systems, like STAT-ON™, try to improve the efficacy of disease management and treatments, and detect the onset of motor complications, and monitor treatment response in the current clinical practice that presents the following major obstacles:

- **Barrier 1: Lack of accuracy and completeness when reporting about own symptoms.** Due to the cognitive impairments, distress, or evasive nature of some of the symptoms caused by PD, the patients often find difficulties or lack sufficient ability to provide reliable/consistent clinically relevant information about the symptoms they experience in order to optimize the treatment. In particular, often the patients are not aware of the onset of dyskinesia and sometimes it is even difficult for them to distinguish between ON/OFF periods. However, these are key information items for the physician to adjust the treatments.

- **Barrier 2: Missing information about the PD symptoms and signs of disease progression at the clinical level.** The currently available means to report and monitor the symptoms are modest compared to the huge challenge posed by the variety of PD symptoms and their fluctuations. The patient's visits and self-reporting may not throw reliable or complete evidence for the physician to cope with the entire picture and overall phenomena surrounding their patients day-to-day. Most of the evidence used builds on reporting provided by the patients themselves and they often lack the ability to undertake this task.

- **Barrier 3: Compromised self-care and adherence to treatments.** Treatment regimens (medications, times, and doses) and adherence to treatment are crucial for correct PD management and for the QoL of the patients. PD patients resort to prescribed regimes, but this seemingly simple commitment may represent a nontrivial feat since patients must add on top of the overall burden the challenge of self-care, which is often difficult to achieve due to the many impairments and distresses linked to the disease. Cognitive deficits such as attention, communication, memory, and executive functions; depression and impulsive behaviors play a key role in the common lack of adherence and self-efficacy in co-management of the disease.

- **Barrier 4: Symptoms recognition in time to better administrate the medication dose.** Another related barrier is the capability of the professional to properly assess the number of OFF hours the patient has experienced to judge, based on that information, the therapeutic effect of the administered therapy since it is based on diaries or patient recall. When an infusion pump therapy is used, the practitioner has difficulties adjusting the infusion rate, and parameters of stimulation, as well as controlling the administration of extra doses.

- **Barrier 5: Usability from the patient's point of view.** Some patients with Parkinson's have OFF phases so severe that they cannot even self-administrate extra doses or medications. Patients with severe OFFs, which have no caregivers who can perform this task for them, often cannot choose the treatment with continuous infusion pumps. So, for these treatment modalities, it is important an accurate evaluation of socio-functional status and resources since support from others is often required.

1.6 Conclusion

As a concluding remark, it could be said that the current knowledge about PD is continually growing, opening the possibilities of new strategies. Many treatments are currently available, requiring a multidisciplinary approach to improve the QoL of the patients. In order to advance in a more personal and patient-centered treatment, the new technologies could help to address new scenarios from a more global perspective, allowing closer and more objective monitoring in real time.

References

[1] GBD 2016 Neurology Global Collaborators. Global regional and national burden of neurological disorders 1990-1996: a systematic analysis for the Global Burden of Disease Study 2016. Lancet Neurol. 2019:18:459–80.

[2] Parkinson disease: a public health approach. Technical brief. Geneva: World Health Organization; 2022. Licence: CC BY-NC-SA 3.0 IGO.: Parkinson disease: a public health approach.

[3] Winter, Y, Balzer-Geldsetzer, M, Spottke, A et al.Longitudinal study of the socioeconomic burden of Parkinson's disease in Germany. Eur. J. Neurol. 2010;17: 1156–1163

[4] Guía oficial de recomendaciones clínicas en la Enfermedad de Parkinson. Matías-Arbelo J coord. Ed SEN. 2019.

[5] Match M, Gerlich C, Ellgring H, Bayés A, et al. Patient education in Parkinson's disease: formative evaluation of a standardized programme in seven European countries. Patient Education and Counseling.2007;65:245–52.

[6] Martínez-Martín P, Calidad de Vida relacionada con la Salud. Ars Médica, 1998.

[7] Forjaz MJ, Frades-Payo B, Martínez-Martín P. The current state of the art concerning quality of life in Parkinson's disease: II. Determining and associated factors. Rev Neurol. 2009; 49:655–60.

[8] Sławek J, Derejko M, Lass P. Factors affecting the quality of life of patients with idiopathic Parkinson's disease--a cross-sectional study in an outpatient clinic attendees. Parkinsonism Relat Disord. 2005; 11:465–8.

[9] Chapuis S, Ouchchane L, Metz O, Gerbaud L, Durif Impact of the motor complications of Parkinson's disease on the quality of life. Mov Disord. 2005; 20:224–30.

[10] Santos García D, de Deus Fonticoba T, Cores C, et al. Predictors of clinically significant quality of life impairment in Parkinson's disease. NPJ Parkinsons Dis. 2021;7:118.

[11] Santos García D, de Deus Fonticoba T, Suárez Castro E, et al. Non-motor symptoms burden, mood, and gait problems are the most significant factors contributing to a poor quality of life in non-demented Parkinson's disease patients: Results from the COPPADIS Study Cohort. Parkinsonism Relat Disord. 2019; 66:151–157.

[12] Pechevis M, Clarke CE, Vieregge P, et al; Trial Study Group. Effects of dyskinesias in Parkinson's disease on quality of life and health-related costs: a prospective European study. Eur J Neurol. 2005; 12:956–63.

[13] Santos García D, De Deus Fonticoba T, Paz González JM, et al. Staging Parkinson's Disease Combining Motor and Nonmotor Symptoms Correlates with Disability and Quality of Life. Parkinsons Dis. 2021;13; 2021:8871549.

[14] Santos-García D, de Deus Fonticoba T, Cores Bartolomé C, et al. COPPADIS Study Group. Predictors of the change in burden, strain, mood, and quality of life among caregivers of Parkinson's disease patients. Int J Geriatr Psychiatry. 2022;37. doi: 10.1002/gps.5761.

[15] Schrag A, Hovris A, Morley D, Quinn N, Jahanshahi M. Caregiver-burden in Parkinson's disease is closely associated with psychiatric symptoms, falls, and disability. Parkinsonism Relat Disord. 2006; 12: 35–41.

[16] Parkinson Study Group. Pramipexole vs levodopa as initial treatment for Parkinson disease: A randomized controlled trial. Parkinson Study Group. JAMA. 2000; 284:1931–8.

[17] Dafsari HS, Martinez-Martin P, Rizos A, et al. EuroInf 2: Subthalamic stimulation, apomorphine, and levodopa infusion in Parkinson's disease. Mov Disord. 2019; 34:353–365.

[18] Frazzitta, G. et al. Intensive Rehabilitation Treatment in Early Parkinson's Disease: A Randomized Pilot Study With a 2-Year Follow-up. Neurorehabilitation and Neural Repair 2015; 29 123–131.

[19] Institute of Medicine (U.S.), Committee on Quality of Health Care in America. Crossing the quality chasm: a new health system for the 21st century. Washington, DC: National Academy Press; 2001.

[20] Gerlach OH, Winogrodzka A, Weber WE. Clinical problems in the hospitalized Parkinson's disease patient: systematic review. Mov Disord. 2011;26:197–208.

[21] Aminoff MJ, Christine CW, Friedman JH, et al. Management of the hospitalized patient with Parkinson's disease: current state of the field and need for guidelines. Parkinsonism Relat Disord 2011; 17(3):139–145.

[22] Van der Eijk M, Faber MJ, Post B, et al. Capturing patients' experiences to change Parkinson's disease care delivery: a multicenter study. J Neurol 2015; 262:2528–2538.

[23] Willis AW, Schootman M, Evanoff BA, Perlmutter JS, Racette BA. Neurologist care in Parkinson disease: a utilization, outcomes, and survival study. Neurology 2011; 77:851–857.

[24] Stocchi F, Bloem BR. Move for change part II: a European survey evaluating the impact of the EPDA Charter for people with Parkinson's disease. Eur J Neurol 2013; 20:461–472.
[25] Dorsey ER, Vlaanderen FP, Engelen L, Kieburtz K, Zhu W, Biglan KM, Faber MJ, Bloem BR. Moving Parkinson care to the home. Mov Disord. 2016;31:1258–62.
[26] Maetzler W, Klucken J and m Horne, M. A Clinical View on the Development of Technology-Based Tools in Managing Parkinson's Disease. Mov Disord. 2016; 31:1263–71.

2

Summary of the REMPARK Project Findings: Innovative Steps

Joan Cabestany

Universitat Politècnica de Catalunya, UPC, Barcelona Spain

Email: joan.cabestany@upc.edu

Abstract

Considering that STAT-ON™ is the final result of some part of the research done in the REMPARK project, this chapter summarizes the contents of the book "*Parkinson's Disease Management through ICT: The REMPARK Approach*," published by River Publishers in 2017, where the research and innovation performed and achieved in this project is described.

Here, an analysis is done about the opportunity associated to the different obtained and reported results. It is explained how a decision was taken on the redesign and additional work done on a wearable sensor for the measurement and detection of movement symptoms related to Parkinson's disease.

The chapter ends with a concrete reference to the initiative for organizing the proposal of an SME Instrument (phases I and II) action project that provided the convenient framework for the development and subsequent launching to the market of a new medical device, commercialized by Sense4Care, with the name STAT-ON™.

2.1 Summary of the REMPARK Project: Objectives, Development, and Findings

It is very well known that Parkinson's disease (PD) is a progressive neurological condition with no cure and only treatments addressed to the management and mitigation of the different symptoms are available to the patients.

In order to contribute to helping patients, the REMPARK project [1] was organized as a challenging initiative with four concrete objectives:

1. The identification of the motor status in real-time conditions. This was supposed to be the identification of the associated relevant parameters and types of motor disorders, and the development of the minimum necessary system, mainly based on wearable inertial sensors and embedded intelligent algorithms.
2. The development of a system for gait improvement, when necessary, based on auditory cueing.
3. The design and implementation of a specific and adapted user interface in order to obtain feedback from the patient using a Smartphone. The application was mainly focused on the interaction with affected people for satisfactory answers to surveys, questionnaires, and prompts generated by the system.
4. The specification and design of a service for the remote management of the disease. The service was planned and based on a server acting as the repository of the whole obtained ambulatory data, in combination with the Electronic Health Record of the patient and making easy the intercommunication between the professionals involved in the caring process and the patients and their relatives when required.

REMPARK was a huge and complex project, with an intensive activity developed from 2011 until 2015, the participation of 11 partners, a well-organized management process, and the involvement of medical institutions and PD patients from four different countries (Spain, Ireland, Italy, and Israel). This was a crucial point for the obtention and construction of the necessary database for the implementation of the machine learning-based algorithms, embedded in the developed sensors for the positive consecution of Objective 1 (the ambulatory evaluation of the critical motor symptoms associated with the disease).

REMPARK project, as can be seen in reference [1], obtained very good results and findings at the end of its execution (see a summary in Table 2.1). A short list of these results is the following:

- A huge and very complete database was obtained. This database was designed according to the objectives of the project and the main interest was to obtain a specific labeled collection of data for the implementation of a supervised learning process in order to get a set of intelligent algorithms to be embedded into the specified sensor prototype (see reference [1], Chapter 4).

2.1 Summary of the REMPARK Project: Objectives, Development, and Findings

The database contains data from 92 patients, obtained from a free activity period at patients' homes, where video recording and annotations were done according to the approved protocol by the Ethical Committees at the participating hospitals from the above-mentioned countries. This activity was covered and done during the first year of the project. At the end of this first experience, REMPARK database contained more than 30 hours of video recording and more than 140 hours of manual annotations of different motor disease symptoms (Dyskinesia, Bradykinesia, Freezing of Gait – FoG), synchronized and correlated to the corresponding tri-axial accelerometers' raw signals obtained from the movement sensor worn by the participating patients.

- A set of algorithmic developments, based on supervised learning methodology using the constructed database and able to identify the specified disease symptoms: Dyskinesia, Bradykinesia, FoG, ON and OFF state estimation, Gait parameters, and fall detection. This algorithmic set (see reference [1], Chapter 4) was designed to be embedded into the sensor prototype (see reference [1], Chapter 5).

- A sensor subsystem prototype able to embed the developed set of intelligent algorithms and to operate in an autonomous way, in order to be used for the implementation of a pilot verification activity, during the last year of the project activity.

- A very important piece for the REMPARK activity development was the patient's Smartphone. The Smartphone was used for interaction with the patients, covering a major part of the communication requirements. It is obvious that the design of an improved and well-adapted user interface was an important task (see reference [1], Chapter 6).

  Apart from other interesting functionalities, the most interesting characteristics for REMPARK purposes were the auditory cueing system controller and the medical questionnaires administration. These questionnaires were automatically sent to the patients, when necessary, after some condition detection not directly related to motor problems.

- The auditory cueing was an actuator REMPARK subsystem able to generate a rhythmic auditory stimulus when some specific condition is detected (gait disturbances, FoG, or Bradykinesia). The details and fundamentals are explained in Reference [1], in Chapter 7.

  In this case, when the condition is correctly detected by the sensor, embedding the developed algorithms, activation of the application implemented on the Smartphone is done. This application is generating a set of rhythmic sounds (auditory cueing) that are sent to an earphone pair worn by the patient.

- A disease management system (DMS) adapted to the specific Parkinson's disease management needs and constraints. Details of this implementation can be found in reference [1], in Chapter 8, where a description of the main organization of the system and its different modules' inter-relationship is done.

  In fact, the DMS system developed for the REMPARK project was an adaptation of an already existing, at that time, generic platform in the Maccabi hospital, in Tel-Aviv (Israel) for the management of other diseases, but with many equivalent needs (storage of data, easy management of the disease, making easy and effective the inter-relationship of the involved professionals, etc.). An important part of this platform was the included Rule Engine module, where knowledge, procedures, and generation of alarms, etc., were included.

During the last year of the project, a validation pilot was organized and completely executed, with the participation of 41 volunteers. The description of the pilot and the main obtained results are presented in Chapter 9 of reference [1]. Remarkable results that must be mentioned are related to the detection of ON and OFF states, as a combination of the also detected movement symptoms. The obtained specificity was 89% and the sensitivity was 98%. The efficacy and effectiveness of the developed cueing system were also measured. Table 2.1 presents a summary of the achieved results in the project, with some annotations and comments on their usefulness for further development and the obtained degree of satisfaction.

2.2 Innovative Technology: Analysis of the Opportunity and Related Challenges

The REMPARK consortium was very happy with the obtained results and findings since the project frame was the opportunity to put in value and take profit of the available technology for more advanced and close care of Parkinson's disease, with the objective of improving the patients' quality of life. Some relevant advances, as indicated in the precedent text, were:

- A version of a wearable prototype able to measure and identify gait and related movement symptoms characterizing Parkinson's disease. Some of the related main advantages are:
  - Only one unique sensor is needed for the detection of all the collection of symptoms.

2.2 Innovative Technology: Analysis of the Opportunity and Related Challenges

Table 2.1 Summary of the main REMPARK project results.

Results	Completeness	Satisfaction with the results	Usefulness for further development	Comments
Labeled database	100% according project specification	√√	Good	To be used for algorithmic refinement and new learning processes
Algorithmic set for motor-related symptoms detection	100% according project specification	√√	Good	Should be embedded in future developments
Sensor prototype	A version was generated for its use in the validation pilot	√	Good	Obtained version arrived at RTL7, and it is able to evolve toward a competitive product
ON and OFF states identification	100% according project specification	√√	Good	Should be embedded in future developments
Cueing system generation	Advanced state	√	Medium	Ideas should be used for future development and experience would be useful for further implementation
Disease management system – DMS	Medium state	√	To be discussed	Not implemented after the project for several reasons

- The wearable is worn at the waist since this is the most appropriate location for correct detection of all the movement-related characteristics in a person.
- The wearable has a large autonomy for continuous use for several days.
- The wearable integrates advanced sensor technology, mainly based on tri-axial accelerometry for gait characteristics measurement.
- The embedded processing capability of the wearable is designed for the complete integration of the developed set of algorithms, converting the device into a really autonomous one with a real-time and on-place processing capacity of the captured data.

• A very complete set of dedicated algorithms, based on AI and learning techniques, for the detection and identification of PD-related movement symptoms (mainly Bradykinesia, Dyskinesia, FoG, and ON/OFF states).

• A very complete system based on a platform for the storage of the captured and processed data, together with a recommender system (the DMS – disease management system) for the implementation of an efficient relationship between the different professionals taking care of the patient. This system was divided into several operative subsystems (the generation of auditive cueing when necessary, sending messages and interaction with the patients and caregivers, through a web-based application) and the most important part was the development of specific applications and interfaces adapted to persons suffering PD on a personal Smartphone.

A very interesting idea and concept was developed during the execution of REMPARK: **the establishment of a double loop of interaction for the enhancement of the quality of life of persons with Parkinson [2]**. A first level was considered and based on a set of wearables and actuators placed at the body's patient level (the worn sensor for the movement analysis and the auditory cueing system), with a high level of autonomy for processing and storing data. A Smartphone was integrated into this patient-level loop in order to facilitate interaction with the patient when necessary and to communicate with the established second-level loop (sending or receiving information).

The second-level loop was considered around a server where data was stored and processed. The core part of this loop is the disease management system module, which should be able to generate recommendations and alerts when necessary, according to the stored and processed information.

A secondary functionality of the second loop of the system was planned as a communication tool between the different professionals taking care of the patients and with the patients or caregivers, when necessary. In fact, the established system is a precursor solution of some actual telehealth initiatives for PD management.

2.2.1 Analysis of the opportunity

REMPARK was a RIA (Research and Innovation Action) in the frame of the EU-WP7 and this means that results, in an optimum case, should arrive at a TRL 7, suggesting that, for real use and transfer of these results, some additional actions are required. Within this context, an exercise of analysis was done, trying to identify which of the findings and results should be good candidates for real development and starting away to the real applicability world and market, if appropriate.

The results and final conclusions of the REMPARK Project were presented in 2015. The incredible evolution of the necessary electronic and sensors' technology, together with the evolving ICT deployment was a good scenario for this analysis, opening the door to a very necessary application in the domain of PD treatment.

As a good exercise, it is possible to follow some published references at that time, in connection with the needs of Parkinson's disease management context and medicine and care, in general [3–5]. An excellent review, published in 2017, can be found in reference [6], giving an overview of the status of the research and the associated challenges in reference to the different symptoms and problems of PD.

Principal ideas and thoughts considered at that moment are the following:

- The wearable prototype developed for the project works and the final pilot deployment and activity was a good example of useful technology for Parkinson's disease management when used for the detection and measurement of the associated disease motor symptoms. The main characteristics of the prototype were:

  - Only one wearable, worn at the waist, is necessary for the detection of movement-associated symptoms.

  - The wearable integrates the whole set of developed algorithms for this detection and measurement of symptoms when processing the related signals generated by the integrated tri-axial accelerometers.

- The processing of this information is done locally, and in real time, and no external connection with a server is necessary.
- The wearable prototype was specially designed for its use at home, and in ambulatory conditions, and for this reason, the autonomy of the integrated batteries is enough for several days.
- The prototype was able to determine the presence and duration of the main motor symptoms associated with the disease, except tremors: Dyskinesia, Bradykinesia, Freezing of Gait, ON and OFF states, gait characteristics, and falls.

In conclusion, **the wearable was considered as a main objective for further development activity and its transformation to a usable and, maybe, commercial medical device.**

- The auditory cueing generation subsystem developed for REMPARK, as an actuator, applicable when gait problems and disturbances are detected, is another example of useful technology considered in this analysis context. The main characteristics of the developed subsystem are:
  - Auditory cueing is generated by an internal application of the associated user's Smartphone.
  - There is a variety of sounds and frequencies available.
  - The application of the cueing can be automatic and decided by the system when gait problems are detected by the worn wearable.
  - Cueing associated sounds are received by the patient using Bluetooth earphones.

The advantages and associated problems to the auditory cueing administration and use for gait problems mitigation in PD are quite well known [7, 8] and are very interesting for further study and applicability problems consideration. Some of the important known and existing problems are:

- The administration of cueing under voluntary activation by the patient could be not as efficient as necessary.
- The most effective type of cueing greatly depends on the specific patient. Not all people are sensitive to the same sounds or frequencies.
- An automatic generation of auditory cueing is complicated since the online and on-time detection of the associated gait problems is not

2.2 Innovative Technology: Analysis of the Opportunity and Related Challenges

completely solved. The sensor developed in REMPARK could be a perfect first step, but the management of the cueing through the Smartphone can suppose a problem, due to the associated delays in the synchronization and application activation processes.

In conclusion, the REMPARK subsystem is a good basis for looking for a correct context in order to advance and obtain further developments and improvements in auditory cueing generation.

- According to Table 2.1, the third relevant result to be considered is the implementation of the disease management system – DMS, adapted to Parkinson's disease needs on a related server. This system is able to store all the information generated during the care process and facilitates a loop of interaction between the different professionals, caregivers, and patients (when necessary). The DMS module, as described in Chapter 8 of reference [1], is able to relate personal data with the Electronic Health Record and to generate alarms, messages, and automatic appointments.

  The system, operated through a web interface application, could be considered a good initiative for the implementation of a convenient telehealth system, adapted to the requirements of PD.

  A telehealth system, including the necessary technology, can be a very good initiative to improve the quality of life of the patients affected by Parkinson. This kind of system can provide an adapted way to conveniently follow the evolution of the patients at home when developing their current activities. It must be seen as a complementary tool to the more traditional visits to the hospital and doctor's office. An interesting presentation of the real possibilities and discussion on the patients' satisfaction is included in the reference [9]. A complementary actual view on that topic, considering the perspectives and advantages, but also the barriers and related problems, can be found in [10]. It is very clear that the implantation of the telehealth service is feasible from the technological point of view, but real barriers still exist, in relation to the required skills for patients and doctors, some existing privacy concerns, the restrictive regulations that are implanted in many countries, and the lack of reimbursement.

Table 2.2 shows a summary of the conclusions obtained during the analysis of opportunities and their related problems.

According to the above-presented text and the Table 2.2 conclusions, a very good opportunity, at the moment when the REMPARK project finished,

Table 2.2 Conclusions of the analysis of opportunity after REMPARK.

Technology	Opportunity – advantages	Related problems	Comments	Feasibility
Sensor	• A good and already working prototype is available. • A complete algorithmic set is embedded in the sensor.	• The sensor autonomy must be increased. • User experience must be improved.	• For its commercialization, the sensor must obtain the qualification of "medical device" and the CE label, according to the European regulations. • Clinical evidence is necessary.	High
Auditory cueing	• Necessary components are considered and included in the prototype.	• Synchronization between the symptom's detection and the launching of the cues is not completely satisfactory. • Detection of the FoG must be improved for a better administration of the cueing.	• Additional initiatives or new projects must be started for satisfactory development and implementation.	Medium
Server-based platform – DMS	• Prototype specification and implementation were satisfactory at the project level.	• In a real implementation, it must be improved communication, data privacy, and security aspects.	• For a real introduction of telehealth in Parkinson's Disease management many aspects must be improved: technology adoption, patient empowerment, doctor's trust, etc.	Low

2.2 Innovative Technology: Analysis of the Opportunity and Related Challenges

was to **concentrate efforts on the final development and possible industrialization of a novel wearable sensor for the detection and measurement of the movement-related symptoms in the mid-stage of Parkinson's disease.** The main ideas and the value analysis of this initiative can be found in the reference [11].

The additional identified technologies (auditory cueing and the server-based platform) were initially discarded for immediate actions, waiting for new opportunities. Among others, the main reasons were:

- For an innovative and effective auditory cueing system, it is necessary to have a good detection device (mainly, able to detect severe movement disturbances and FoG episodes in real time) and able to automatically launch the generation of the auditory cueing.

- The server-based platform is, in essence, a very good idea. It is a solid step through the consolidation of the main eHealth ideas and initiatives. At that time, we considered that it was too early since the development of the necessary interoperability and sharing measures of personal data was still very country dependent.

In this way, the REMPARK partners owning the IPR of the family of AI-based algorithms to be included in a possible manufacturable sensor device decided to go away and start a series of necessary actions for the materialization of this idea.

2.2.2 Related challenges

Once it was decided to progress for the obtention of a new wearable sensor device for helping the PD community in the detection, measurement, and following of the disease evolution, it was necessary to face some important points:

- To determine the best development context. It seems that research or innovative action is not the correct environment to get a final product, ready to be launched to the market and society.

- To decide which is the correct final product format, compatible with the above ideas. The initial thought was to go to the materialization of a medical device, but this must be refined.

- To determine the most advantageous organization to correctly advance in these objectives. The main question is about the suitability of a Spanish university (Universitat Politècnica de Catalunya) as the main owner of the IPR, for covering all the necessary stages to cover the objective in an agile way.

- To obtain resources and enough funding to cover the complete initiative.

After a complete analysis of the needs and opportunities, it was decided that the most efficient way to work was under the modality of SME (Small and Medium Enterprise), trying to get resources from the available EU actions at that moment (SME Instruments action).

With this idea, Universitat Politècnica de Catalunya (UPC) was signing a technology transfer contract, for the commercial exploitation of the IPR related to REMPARK results, with Sense4Care SL, an SME company created in 2012, and participated by the UPC. Part of the UPC researchers, taking part in REMPARK, were co-founders and owners of this company.

Sense4Care was proposing the PARK-IT project (*Unobtrusive, continuous and quantitative assessment of Parkinson's disease: hard evidence for optimal disease management with information technologies*), which was granted with an SME Instruments – phase I action, under contract number 672228 in 2015. With this funding, Sense4Care was able to study and take conclusions about the market opportunity, and associated business model for the presented initiative.

2.3 The PARK-IT Project: Main Conclusions

The aim of the phase 1 project, called PARK-IT, was to confirm the feasibility of a successful launching to the market of the future product. As the technical aspects of the sensor were verified during the REMPARK project, achieving over 90% sensitivity and over 90% specificity, the key for the phase 1 project was to confirm that there is sufficient market demand for the PARK-IT product. It was distributed into three different tasks:

- Market and stakeholder analysis: in order to determine the key market and its size, market drivers, and routes to market and to identify the key stakeholders in order to establish possible partnerships with them.

- Development of a detailed Plan for regulatory aspects and IP Management, including a CE medical device certification plan.

- Drafting an elaborated business plan, including target and value proposition, possible distribution channels and price strategy, manufacturing cost, gross margin, marketing strategy, and the establishment of the company structure and organization...

2.3 The PARK-IT Project: Main Conclusions

Sense4Care was developing the PARK-IT work along the project's scheduled time, analyzing the market opportunities and concrete needs, and identifying the possible users of the proposed solution.

An important conclusion was that the PARK-IT product can be considered a **Class IIa medical device** and a detailed plan for regulatory aspects and IP management was elaborated. As a final conclusion, what this feasibility study was demonstrating is that, once the project is completed, PARK-IT should be the market-leading solution, with a significant target market and a clear business strategy to achieve a successful market launch.

The conclusion of the action was that Sense4Care should submit an application to phase 2 of the SME Instrument action, in order to obtain the necessary budget for the required work to do.

2.3.1 The following steps

After the conclusions obtained from the PARK-IT project, it was time for the preparation, study, and edition of a new project proposal, with the same title and the PARK-IT2 acronym, that was submitted to the SME Instruments – phase 2 action. PARK-IT2 received the necessary support and funding, starting the associated works at the beginning of 2017, under contract number 756861. This project, with a scheduled duration of 24 months, was organized around the following work packages:

- WP1. Redesign of PARK-IT in order to obtain a version "ready-to-market." The estimated duration was 14 months, starting at the beginning of the project.

- WP2. CE medical device certification. The estimated duration was 16 months, starting when the redesign should be in an advanced state, and finishing at the end of the project.

- WP3. PARK-IT demonstration pilot. The estimated duration was 11 months, to be started with the newly redesigned prototype would be available and finish at the end of the project.

- WP4, WP5, and WP6 have a transversal character and were scheduled throughout the whole project duration. They must cover: the communication and dissemination parts, the IPR and commercialization-business strategies, and the global management of the project.

During the execution, it was decided to add a new and necessary WP7 on "Ethics requirements and protection of personal data." Finally, after a

Figure 2.1 Redesigned PARK-IT wearable prototype.

proposed and accepted Amendment to the contract, the duration of PARK-IT2 was enlarged by 6 months, concluding in a satisfactory way on October 2019.

The completely redesigned wearable sensor during the project PARK-IT2 project (shown in Figure 2.1) was the prototype originating the actual medical device, commercialized by Sense4Care, with the STAT-ON™ registered name.

2.4 Conclusion

A summary of the findings and technological advances found in the REMPARK project development are presented. An analysis of opportunity is done, motivating the proposal and work done in the frame of the PARK-IT2 initiative, which generated the precursor of the current STAT-ON™ medical device.

References

[1] J. Cabestany, A. Bayes (editors), 'Parkinson's Disease management through ICT: The REMPARK Approach' River Publishers series in Biomedical Engineering, ISBN 978-87-93519-45-9 (Ebook), 2017

[2] A. Samà, C. Pérez-López et al. 'A double closed loop to enhance the quality of life of Parkinson's Disease patients: REMPARK system' Studies Health Technol. Inform. (2014) 207:115–124. IOSPress doi: 10.3233/978-1-61499-474-9-115

[3] W. Maetzler, L. Rochester. 'Body-Worn sensors – the Brave New World of Clinical Measurement?' Mov. Disorders (2015) 30: 1203–1205 doi: 10.1002/mds.26317

[4] A. J. Espay, P. Bonato et al. 'Technology in Parkinson's Disease: Challenges and Opportunities' Mov.Disorders (2016) 31: 1272–1282 doi: 10.1002/mds.26642

[5] J. M. T. van Uem, T. Isaacs et al. 'A Viewpoint on wearable technology-enabled measurement of wellbeing and health-related quality of life in Parkinson's Disease' Journal of Pak. Disease (2016) 6: 279–287 doi: 10.3233/JPD-150740

[6] E. Rovini et al. 'How wearable sensors can support Parkinson's Disease diagnosis and treatment: A systematic review' Frontiers in Neuroscience (2017) 11: article 555 doi: 10.3389/fnins.2017.00555

[7] W.R.Young et al. 'Auditory cueing in Parkinson's patients with freezing of gait. What matters most: Action-relevance or cue-continuity?' Neuropsychologia (2016) 87:54–62 doi: 10.1016/j.neuropsychologia.2016.04.034

[8] J. Gómez-Gonzales et al. 'Effects of auditory cues on gait initiation and turning in patients with Parkinson's disease' Neurologia (2019) 34 (6): 396–407

[9] J. R. Wilkinson et al. 'High patient satisfaction with telehealth in Parkinson disease' (2016) Neurology: Clinical practice 6: 241–251

[10] A. Shalash et al. 'Global perspective on Telemedicine for Parkinson's Disease' Journal of Park. Disease (2021) 11: S11–S18 doi: 10.3233/JPD-202411

[11] A. Bayes et al. 'A "HOLTER" for Parkinson's disease: Validation of the ability to detect on-off states using the REMPARK system' Gait & Posture (2018) 59: 1–6 doi: 10.1016/j.gaitpost.2017.09.031

3

The STAT-ON™ Industrialization Pathway: From the Research Prototype to the Product

Carlos Pérez López[1], Daniel Rodríguez-Martín[2], and Martí Pie[2]

[1]CSAPG – Consorci Sanitari de l'Alt Penedès i Garraf. Research Department, Spain
[2]Sense4Care SL – Cornellà de Llobregat, Spain

Email: cperezl@csapg.cat; (daniel.rodriguez) (marti.pie)@sense4care.com

Abstract

After the technical verification of the REMPARK sensor for the detection and measurement of the motor symptoms associated with Parkinson's disease (PD), a complete redesign process was necessary to obtain a version with feasible industrialization characteristics.

The chapter presents the followed process, with many details about the established final requirements of the system, how the new internal architecture was organized, and how the necessary embedded firmware was implemented.

Details on the mechanical design and the final packaging and labeling of the STAT-ON™ device are included.

3.1 Introduction

The industrialization of a medical device assumes the capacity to organize serial production of a specific device compliant with a list of required and specific standards. In our case, STAT-ON™ is based on the knowledge and technology generated in several Spanish and European-funded research projects such as REMPARK [1,2] and PARK-IT phase I, among others, as has been mentioned in chapter 2.

The industrialization of a medical device must be according to the standard IEC60601-1 for medical electrical equipment. Thus, some requirements must be defined and specific electrical circuits must be designed accordingly. Additionally, the elements that surround the device, such as the sealing strip, enclosure, packaging, and the accompanying documents, must be defined at the beginning of the project.

In parallel, and as explained in Chapter 4, this process must be aligned with the complete certification process. In the present chapter, it will be described the industrialization process of STAT-ON™, from the prototype to the final product.

The main achievement of the REMPARK project was the development of an algorithmic set capable of detecting Parkinson's Disease (PD) motor symptoms with a sensitivity and specificity greater than 85% for all symptoms of interest. These algorithms, based on machine learning strategies combined with frequency and statistical analysis on inertial signals, also include adaptive techniques that allow to adapt the outcomes to the PD patient profile.

As a continuation of the REMPARK project, a Feasibility Study (PARK-IT phase I) was carried out on the potential market for a medical device to monitor patients with PD, to understand if there was enough demand. The conclusion of this study was positive and indicated, at the same time, that many changes would be necessary to the REMPARK prototype to adapt the device to the market demand.

Originally, the REMPARK proposal assumed a connection between the sensor and a hospital platform, where the data was stored. Clinical professionals could connect to the platform to analyze the data, monitor the patient and manage the concrete treatment. This architecture, at that time, was not feasible for a commercial solution, such as the one proposed, due to the existent interoperability issues among the different hospital systems and the certifications/specifications necessary to connect to them.

For this reason, in the PARK-IT study, a stand-alone system architecture was proposed, where the clinical professional will manage the sensor through a portable device (Smartphone) and where the patient's data is always under the responsibility (in custody), either of the patient himself or of the clinician in charge of the patient care.

After the REMPARK pilot experience and along the Phase I feasibility project, it was concluded that there were two essential aspects to improve on the original sensor: (1) the duration of the monitoring must expand from a few hours to a complete week and (2) the usability of the sensor and its charging system should be improved, when compared to the initial prototype, allowing the patient to wear it more comfortably.

As it is presented in Section 3.1 of Chapter 2, an industrialization process began (PARK-IT2 initiative) that included a general redesign of the initial research prototype towards a medical device, capable of sustaining a feasible business model.

3.2 The Requirements of the STAT-ON™ System

In the PARK-IT phase I feasibility study, the following sensor use scenario was proposed: *the sensor will be under the responsibility and must be configured by the neurologist since the algorithms need a concrete number of clinical data from the patient to adjust the detection algorithms. Once the sensor is configured, it would be delivered to the patient, who would wear it for at least 4 days.*

Monitoring can be extended for as long as the neurologist considers necessary, taking into account that, if necessary, the patient could charge the battery of the sensor at home overnight. Once the monitoring will be finished, the patient or caregiver should send back the device to the neurologist's office, where the sensor's data would be downloaded and, automatically, a complete monitoring report would be generated using a developed Smartphone app.

In this way, the **sensor was conceived as a small portable device that, worn on the left side of the waist, is capable of continuously monitoring and evaluating the PD patient's motor symptoms**. The following requirements are crucial for the presented device concept:

- It must be based on inertial sensors and should have the ability to process data in real time.

- It would include a rechargeable battery providing complete autonomy for several days.

- The data, once processed, would be stored in the internal memory of the device.

- In addition, the sensor must have the ability to communicate wirelessly and safely with a Smartphone and transfer the stored data in a secure way.

- The hardware device must include enough computing capability to embed the signal-processing algorithms and methods developed during the REMPARK project.

The additional pillar of the system should be a mobile application for the generation of complete reports and time distribution of the symptoms in a useful

Figure 3.1 STAT-ON™ data flow scheme (patient is wearing the sensor for a period of 7 days while doing normal activities).

format for neurologists. In fact, this application would act as the system's user interface, allowing the neurologist to know the status of the sensor and download the reports generated from the patient's monitoring.

The presented scenario is sketched in Figure 3.1 and the more complete list of requirements is in Table 3.1. According to it, and from a technical perspective, the more challenging aspect to be considered is the reduction of the global consumption of energy, since the original REMPARK prototype had a very high one, due to the main microprocessor used that was the response of many of the internal operations (control aspects and on-line running of the algorithmic set).

The strategy for the consumption reduction was addressed with the use of two different microprocessors, sharing the execution of the different internal processes: the Nordic nRF51822 (nRF), a low-consumption device that will be used for the operational control part of the sensor [3], and the STM32F415 (ST), a high-performance microprocessor that will be responsible for the execution of the algorithmic set in real time [4].

This strategy completely changed the functional scheme of the sensor, evolving towards a hierarchical structure of two different microprocessors. This architecture allowed, on the one hand, to stop the high-performance processor (ST) at convenience when no movement was detected or when the minimum operating conditions were not met. On the other hand, it also allowed the nRF processor to turn on the algorithmic execution, turning on the ST, when necessary.

This scheme requires that each microprocessor must have an independent accelerometer connected. The accelerometer connected to the nRF would allow the data capture and analysis process to be started by turning

Table 3.1 Complete list of the sensor requirements.

Requirements Type	Sensor device requirement description
Functional	The DEVICE must wake up when "wake-up situation" is detected (trembling).
	The DEVICE must start collecting data after "wake-up situation" and save it in a Flash memory.
	The DEVICE must store the push button events.
Power management	The DEVICE must have an autonomy of at least 7 days with normal use.
	The DEVICE's battery must charge with wireless power system (Qi standard).
	The DEVICE must be fully charged in less than 6 hours.
	The DEVICE must be able to detect low battery status through a battery gauge.
Communication	The DEVICE must be able to send and receive data using BLE (Bluetooth low Energy).
	The DEVICE must be able to send the saved data from the Flash memory to the APP (from last synchronized time stamp).
	The DEVICE must be configurable from the APP to synchronize key parameters and alarms.
Human machine interface (HMI)	The DEVICE must have a RGB LED and a monochromatic ORANGE LED.
	The ORANGE LED will display information about the status of the charge of the battery (controlled by the charging circuit).
	The DEVICE will turn on RGB_BLUE LED when connected through Bluetooth. After it will return to GREEN or WHITE.
	The DEVICE will blink RGB_WHITE LED when it is not yet configured. After it will return to GREEN or OFF.
	The DEVICE must have a push button. After the button is pushed acoustic or vibration feedback will be activated.
	The DEVICE must have a buzzer. The buzzer may be replaced by a vibrator if space and the device's autonomy is not affected.
Mechanical	The DEVICE's housing must be IP 65.
	The DEVICE's housing must be ergonomic.
	The DEVICE is fixed onto a BELT.
	The BELT must secure the DEVICE to the hip.

on the ST. The ST microprocessor would use its own accelerometers to capture data and perform the relevant analysis, executing the algorithmic core. When the ST detects a prolonged absence of movement, all internal processes will stop and a request will be sent to the nRF to cut off the power supply.

The complete development of the specified sensor device will suppose, at least, the implementation of the following components:

- An electronic board containing and interconnecting all the required electronic components, with the necessary firmware, to implement the sensor monitoring and the processing and communication functionalities.
- An IP 65 housing to contain the mounted electronic board.
- A resistive keyboard to implement the human-machine interface.
- A belt to attach the sensor device to the human body, in the waist.

The following sections describe the development work done to achieve the mentioned components. Details can be found on hardware electronics, firmware, and mechanical design.

3.3 The STAT-ON™ Hardware Electronics

The application of the above-mentioned requirements in the STAT-ON™ redesign process resulted in a quite complex system with a mandatory reduction of consumption, the wireless battery charge possibility, the use of a microprocessor with enough capability to compute machine learning algorithms with a floating-point unit, the inclusion of a microprocessor with Bluetooth capabilities and low-power capacities to perform the operational control part, the implementation of the necessary human machine interface (HMI), that needed a series of communication buses for its efficient implementation...

The present section describes the electronic circuitry of STAT-ON™, how it was industrialized, and how the design was performed to reduce electromagnetic compatibility problems. It includes a discussion on the main circuit architecture, including the power system as a crucial part of this redesign, the chosen inertial sensors, and how they are used, and the section will finish with the presentation of the designed printed circuit board (PCB).

3.3.1 The main circuit

The complete system is composed of three different subsystems (see Figure 3.2):

- One part, based on the ST microprocessor, is in charge of acquiring the main data and processing it in real time (in yellow).

3.3 The STAT-ON™ Hardware Electronics

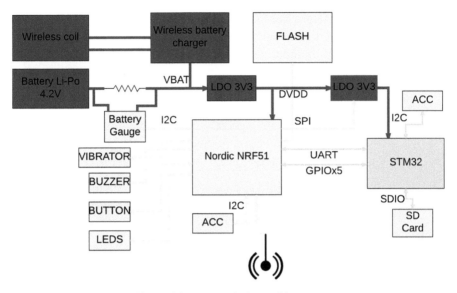

Figure 3.2 Sensor device architecture.

- The second part is based on the nRF microcontroller, in charge of controlling the complete system and sending the data through a Bluetooth communication link (in blue).
- The third subsystem is the power management one to supply all parts of the circuit and the battery (in red).

The chosen ST microprocessor for the online data processing was already used in the REMPARK prototype and this decision facilitated the migration of the developed algorithms to the new platform. This microprocessor will be in charge of reading the accelerometer sensor data, processing them by the embedded algorithmic core developed in REMPARK, storing them in the SD card, and sending them to the nRF microcontroller, when necessary.

The ST microprocessor is part of the STM32F415xx family and is based on the high-performance ARM® Cortex®-M4 32-bit RISC core, operating at a frequency of up to 168 MHz. The Cortex-M4 core features a floating-point unit (FPU) single precision which supports all ARM single-precision data-processing instructions and data types. It also implements a full set of DSP instructions and a memory protection unit (MPU) which enhances application security. In concrete, the STM32F415RGT6 incorporates high-speed embedded memories with 1 Mbyte of Flash and 192 Kbytes of SRAM, and an extensive range of enhanced I/O. This model offers three

12-bit ADCs, two DACs, a low-power RTC, 12 general-purpose 16-bit timers including two PWM timers for motor control, two general-purpose 32-bit timers, a true random number generator (RNG), and a cryptographic acceleration cell. For our redesign, their included communication interfaces capabilities are of special interest:

- up to three I2Cs,
- three SPIs,
- four USARTs plus two UARTs,
- an USB OTG full-speed and a USB OTG high-speed with the full-speed capability, and
- an SDIO/MMC interface (for SD card).

From a more general point of view, this microprocessor's family operates in the temperature range –40 to +105 °C, from a 1.8 to 3.6 V power supply. The supply voltage can drop to 1.7 V when the device operates in the 0 to 70 °C temperature range. A comprehensive set of power-saving modes allows the design of low-power applications.

The chosen microprocessor to manage the whole system and establish a connection to external devices (using Bluetooth 4 protocol) is an nRF51822 from Nordic [3]. This microcontroller is a powerful multiprotocol single-chip solution for Ultra Low Power (ULP) wireless applications. It incorporates Nordic's latest best-in-class performance radio transceiver, a 32-bit ARM® Cortex™ M0 CPU, and 256kB/128kB flash and 32kB RAM memory. The nRF51822 supports Bluetooth® low energy (formerly known as Bluetooth Smart) and 2.4 GHz protocol stacks.

The Programmable Peripheral Interconnect (PPI) system provides a 16-channel bus for direct and autonomous system peripheral communication without CPU intervention. This brings predictable latency times for peripheral-to-peripheral interaction and power-saving benefits when the CPU is in an idle state.

It is interesting to note that the nRF microcontroller has two global power modes ON/OFF, but permitting individual power management and control for all the internal blocks and peripherals. This allows the designer to switch RUN/IDLE the system blocks based on specific requirements derived from particular tasks. These characteristics' set is the basic reason for choosing this microcontroller for the redesign process.

In relation to the power management part (the red part of Figure 3.2), a 3.3V constant voltage is used to supply the nRF through Low Dropout

Regulators (LDO). A second LDO is used as a power switch to supply the ST microprocessor and implement an active low-power strategy (the ST microprocessor is only supplied when necessary).

The Nordic nRF sub-system is composed of the following devices (blue part of Figure 3.2):

- A wake-up accelerometer so the nRF (and complete system) is awoken when movement is detected in the device.
- Flash memory to store all the required data received from the ST or the external App.
- A battery gauge to monitor the battery level.
- A conditioning input circuit to detect when an external button is pushed.
- An enabling circuit to activate the buzzer and vibrator.
- A circuit to control the LEDs.

The nRF microcontroller communicates with the ST microprocessor through a UART port along with two signals that allow proper implementation of an interruption-based communication. Finally, the ST sub-system is composed of the (yellow part in Figure 3.2):

- An accelerometer to read all movements detected by the sensor, when it is awake.
- A slot for flash memory to store all the generated data from the algorithms implemented and executed in the microprocessor. This memory has an additional debugging purpose.

3.3.2 The power system

As mentioned before, the power system is a critical part of the STAT-ON™ redesign. It has been designed considering that it must supply three different parts, with different characteristics: the digital and analog parts corresponding to both microcontrollers (nRF and ST) and the power supply system.

The three systems are separated by ferrite beads to prevent electromagnetic interferences (EMI) and the voltage level is established by means of LDO. Additional ferrite beads have been included between the regulators and the different loads as a preventive measure.

The control and stabilization of the voltage in the inputs and outputs of the included regulators is done, as usual, by capacitors. These controls are

46 The STAT-ON™ Industrialization Pathway

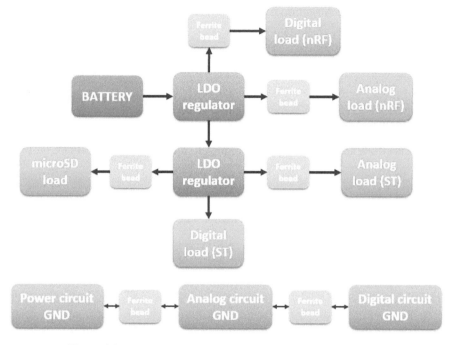

Figure 3.3 The power system and regulator management scheme.

very useful for highly demanding components, like the used microcontrollers, flash memory, or SD Card.

Figure 3.3 shows the main scheme of the STAT-ON™ power system, separated by the different zones.

As shown in Table 3.1, the STAT-ON™ must be a wirelessly charged device according to the Qi standard, and more concretely, with WPC v1.1 Qi Industry Standard. Following the strict requirements of this standard, the redesign must guarantee that the device will work correctly and will not have problems with electromagnetic emissions or other external agents.

As the wireless power charger, the BQ51050BRHLR from Texas Instruments was chosen due to its integrated dual functionality: Qi-receiver and battery charger (integrates the digital controller required to comply with Qi v1.2 communication protocol, and provides all necessary control algorithms needed for efficient and safe Li-Ion and Li-Pol battery charging). The wireless charger system is completely autonomous, starts a charging cycle if the battery voltage is above a threshold, and detects if the device is placed on a charging platform.

For the design of load modulation capacitors, which are in charge of the correct communication between the emitter and receiver, the instructions provided by Texas Instruments were strictly followed. A specific design was done for STAT-ON™ system to test and adjust the correct values for the capacitors and resistors as it is shown in Figure 3.4.

The battery is a fundamental internal element of the sensor redesign. It should be selected maintaining a balance between two essential design restrictions:

- On the one hand, its capacity must be maximum, to increase the autonomy of the sensor.

- On the other hand, the size of the global system must be restrained as much as possible to increase user comfort.

A lithium polymer battery was selected, according to the specifications in Table 3.2.

The selected battery is a lithium polymer battery with a size of 50 × 34 × 6.5 mm^3 and a weight of 24 g, which guarantees safety, it is compliant with all the related regulations, and its size is correct for the designed enclosure. This battery is certified with different standards, specially IEC62133 which is specific for medical device purposes. It includes a short-circuit protection circuit and power is cut under 2.7 V to avoid malfunctioning of the battery and dangerous conditions. The working temperature range and the charging mode must be carefully considered since there exist some strict limitations: from 10 to 45°C when in charging mode and from 10 to 60°C when in discharging mode.

The battery gauge is a device included in the system (see Figure 3.2) to provide information about the battery capacity and its state of charge, using internal processing algorithms. The device will send the information to the nRF microcontroller.

The chosen battery gauge is the BQ27441DRZR-G1 from Texas Instruments. This device includes a patented embedded algorithm to estimate the battery capacity. This device has an I2C bus for configuration and information purposes.

The temperature control of the battery is essential and mandatory for a medical device. Modes of charge not only depend on the phase of charge but also directly depend on the temperature. The used power charger (BQ51050B) is JEITA standard compliant and this implies a series of rules to charge the battery in a safe way: current and voltage limitations depending on the temperature (details are in Figure 3.5). Nevertheless, the battery manufacturer strongly recommends not to charge the battery over 45°C, and this is the reason why, in the redesign process, we do not allow the battery to charge over 45°C.

48 The STAT-ON™ Industrialization Pathway

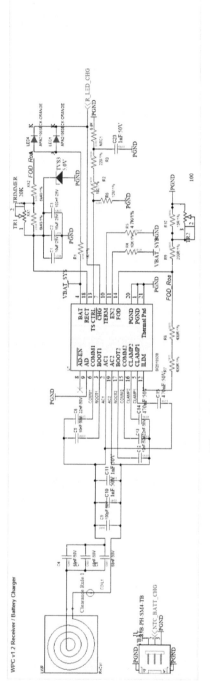

Figure 3.4 WPC V1.2 receiver power system.

3.3 The STAT-ON™ Hardware Electronics

Table 3.2 Battery specifications.

Name	Lithium-ion polymer battery
Voltage	3.7 V
Capacity	1150 mAh
Dimensions ($L \times W \times T$) mm	50 × 34 × 6.0
Standard charge current	0.2/0.5C
Maximum charge current	1A
Open circuit voltage	3.7–3.9 V
Cut off voltage	2.75/4.2 V
Cycle life	500 times
Appearance	Without scratch, distortion, contamination, and leakage
Certifications	CE, RoHS Directive-compliant, UL, SGS, BIS, CB, IEC 62133

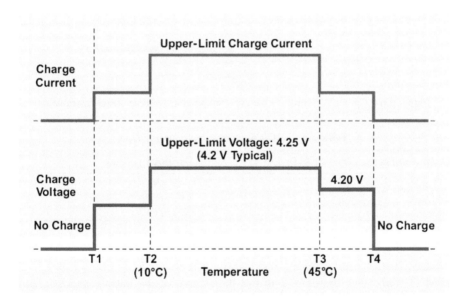

Figure 3.5 JEITA guidelines for charging Li-ion batteries (notebook applications).

To implement the control of the battery temperature and according to the BQ51050B datasheet, the circuit of Figure 3.6 is used. It is necessary to calculate resistors R1 and R3 according to a specific range of temperatures and regarding a specific NTC.

Following the equations on the BQ51050B datasheet, a thermistor with $R_0 = 6.8$ KΩ and $K = 4480$, and using the application provided by the

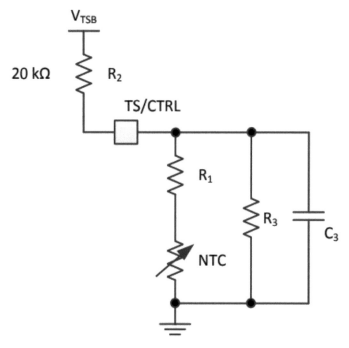

Figure 3.6 Circuit to control the battery's temperature. Source: Texas Instruments.

Table 3.3 LDO's dropout voltage specification.

Output current, $T_j = -40°C$ to $125°C$	Dropout voltage
1 mA	5 mV
200 mA	133 mV
300 mA	200 mV

manufacturer to compute R_1 and R_3 (BQ5105XB NTC Calculator Tool, (SLUS629)), we determined the following values: $R_1 = 400\ \Omega$ and $R_3 = 10M\ \Omega$. This way, the device allows the charge of the battery process from −0.58°C to 45.4°C, being the used thermistor an NT, Serie B57371V2, which is compatible with the Qi charger and the used battery.

The chosen LDO is an LP3981IMM-3.3/NOPB from Texas Instruments, able to provide a maximum output current of 300 mA. This LDO has been chosen due to its ultra-low voltage dropout (to maximize battery capacity) (see Table 3.3).

From current consumption measurements, it is possible to observe that the maximum current will be 200 mA and the voltage dropout will be around 130 mV. This LDO has an *Enable* pin to turn the LDO regulator off. This pin is used to disconnect the ST microprocessor during some operation modes when the nRF performs the control and management of the system.

3.3.3 The inertial sensors

According to the specifications, the system will use two accelerometers: one connected to the nRF and the other to the ST.

The chosen accelerometer is a LIS3DHTR from an STMicroelectronics manufacturer. This accelerometer has dynamically user-selectable full scales of ±2G/±4G/±8G/±16G and is capable of measuring accelerations with output data rates from 1 Hz to 5.3 kHz. The self-test capability allows the user to check the functioning of the sensor in the final application. The device may be configured to generate interrupt signals using two independent inertial wake-up/free-fall events as well as by the position of the device itself.

The LIS3DH has an integrated 32-level "first-in-first-out" (FIFO) buffer, allowing the user to store data to limit the intervention of the host processor. The LIS3DH is available in a small thin plastic land grid array (LGA) package and has a guaranteed operation over an extended temperature range (from -40 °C to +85 °C). The internal embedded registers may be accessed through both the I2C and SPI serial interfaces.

This flexibility and the rest of exhibited characteristics motivated the selection of this accelerometer for its use in this redesign:

- The automatic detection of events and the ability to generate interrupts make it an ideal option for the functionality associated with the nRF microcontroller.
- The option of using a FIFO to store data autonomously, the great variety of selectable full scales, and the possible sampling frequencies make it an ideal option for the continuous data capture to be processed by the ST microprocessor.

In addition, the possibility of using two different serial communication interfaces (I²C or SPI), allows us to use the most convenient one, depending on the data transfer requirements.

3.3.4 The printed circuit board (PCB)

The final version of the PCB is based on the design done for the REMPARK prototype. Several modifications were introduced to guarantee the correct

52 The STAT-ON™ Industrialization Pathway

Figure 3.7 Antenna connection and isolation using vias connected to ground.

functionality, reduce the production cost by setting a strict "design rules" requirement, and delete redundant or unused components. Some of the most relevant characteristics of the final industrialized PCB version are:

- The board has two layers, top and bottom. All the components have been included in the top layer to cheapen the manufacturing process in series production.
- All the power lines must have a width of 0.5 mm maximum and 0.2 mm minimum, while the signal lines cannot be wider than 0.3 mm and the preferred width is 0.2 mm.
- Vias must be 0.55 mm width in diameter with a hole of 0.2 mm minimum.
- Clearance between tracks must be 0.15 mm minimum.
- Vias have a direct connection to obtain a better distribution of the current between the top and bottom layers.
- The antenna must be isolated by a line of vias connected to the ground as the Bluetooth BLE requirements suggest (see Figure 3.7).

Another aspect to take into consideration is the isolation of the critical and vulnerable zones in the PCB, such as the crystal clocks, which oscillate to several Mhz, since the design must secure the electromagnetic isolation to not interfere the general clocks of the microcontrollers. A specific ring has been included for the improvement of the clocks' performance, as can be seen in the Figure 3.8.

The final PCB, as mentioned, has been organized into two layers, with all the components mounted in the top layer for making the assembly process cheaper. The shape of the PCB has been designed to correctly fit and adapt to the enclosure, also providing enough space for the battery. Short connection wires are used between the PCB and the battery, the coil of the charger, and the membrane button. See the final aspect in Figure 3.9.

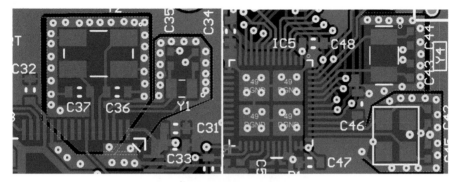

Figure 3.8 Crystal circuit rings.

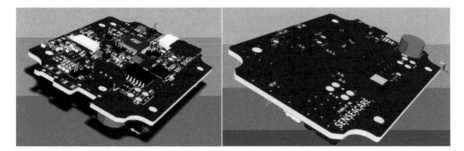

Figure 3.9 3D circuit model views.

3.4 The STAT-ON™ Firmware

As specified above, the STAT-ON™ architecture includes two different processors:

- the nRF microcontroller for the general system control and wireless communication with the external user's Smartphone device, and
- the ST microprocessor for acquiring data from the accelerometer and processing them in real-time for obtaining relevant information about the PD symptoms.

The present section describes the implemented firmware for both processors, including, the implemented protocol for data transfer between both, when required.

3.4.1 Firmware for the Nordic nRF51822

The concrete Nordic microcontroller used is the nRF51822 (nRF). To implement the required and scheduled functionality, it must manage a number of

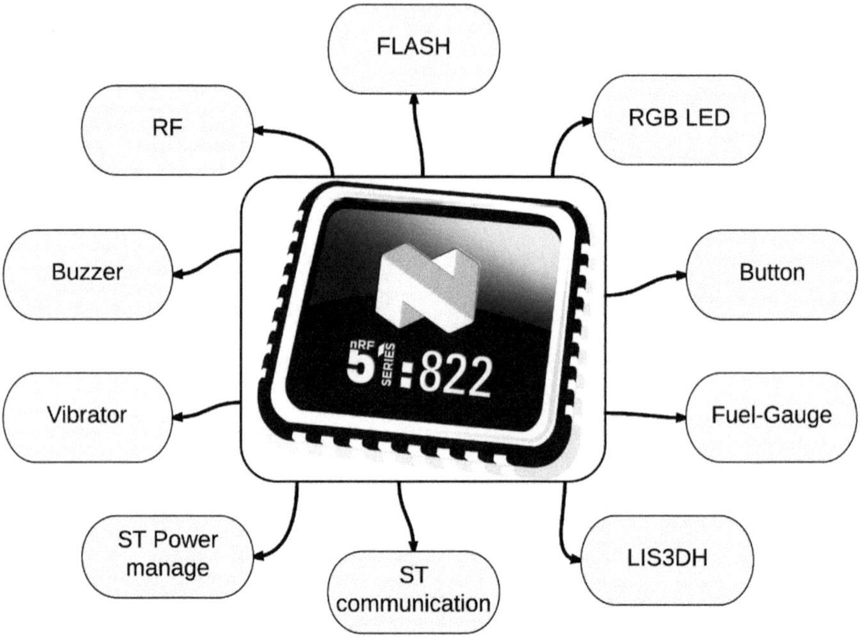

54   *The STAT-ON™ Industrialization Pathway*

Figure 3.10   nRF51822 system with related modules.

modules (the human interface with the machine (HMI), the internal interconnection with the ST microprocessor, and the management of the Flash memory...). An overview of the nRF-related modules is shown in Figure 3.10.

A brief explanation of the purpose of each module is following:

- RF: It is the block to manage wireless communication, which in this case it is Bluetooth low energy (BLE).
- Buzzer: The buzzer is used for notifications and/or alarms to the user.
- Vibrator: The vibrator is used for notifications and/or alarms to the user.
- ST power management: The nRF is able to switch ON/OFF the ST microprocessor. This module is very useful for minimizing power consumption.
- ST communication: A protocol for serial communication, using UART, between the two microcontrollers was developed from scratch.
- LIS3DH: For the management of the used accelerometer.
- Fuel-Gauge: This module provides the possibility to be aware of the battery level in a very precise way.

3.4 The STAT-ON™ Firmware

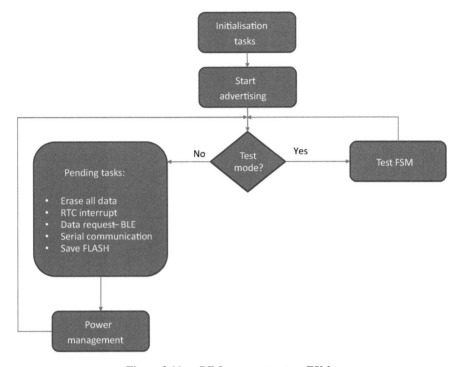

Figure 3.11 nRF firmware structure FSM.

- Button: Provides the management of the mounted button.
- RGB LED: Provides the management of an RGB LED.
- FLASH: For the management of the Flash memory.

3.4.1.1 Main structure of the nRF firmware

The nRF firmware is implemented following the structure and organization of a typical Finite State Machine (FSM), as indicated in Figure 3.11, and according to the following functionality:

Initialization Tasks Block
This block is in charge of the correct initialization of all the peripherals in the system and the correct assignment of the initial values to the variables used in the program. The different tasks executed are:
- **Clock**: It is to notify the nRF that a 32 MHz crystal is being used, start the external high-frequency crystal and wait for its stabilization.

- **Alarms**: Selection of the default mode of the alarms (vibrator and two beeps). The counter's variable for the number of beeps within an alarm is reset.
- **GPIOs**: Configuration of all the general-purpose input/output pins used in this application that are not configured on the particular initializations' blocks.
- **ST interrupt pins**: Interrupt and handling capabilities configuration of the pins used by the ST microprocessor to notify the nRF about new data or no more data. The interrupt is set every time one of these pins changes its state from low to high.
- **Flags**: Clear all the used flags.
- **Data management**: It resets the circular buffer used in the serial communication between ST and nRF microcontrollers.
- **Timers create**: Three different timers for the correct management of the Real Time Counter (RTC), the timeout of the serial communication, and the alarms are created.
- **Buttons and LEDs**: Initialization of the button and the RGB LED.
- **BLE functions**: All this functionality is provided by the manufacturer and the purpose of this task is to initialize the GATT table, the stack, and other parameters regarding the Bluetooth LE protocol and connection parameters/states.
- **Timer starts**: Only the RTC timer is started.
- **LIS3DH**: Configuration, via the serial port I2C, of the accelerometer to generate an interrupt when movement is detected.
- **Fuel-gauge**: Configuration, via I2C, of the fuel-gauge device BQ24771 for the used battery.
- **FLASH**: Initialization of the flash memory.

Start advertising block
Once the initialization process is finished, the device starts advertising via Bluetooth. This makes the device discoverable and connectable.

Test mode block
The program checks whether the device is in test mode (for technical issues) or in normal operation mode.

3.4 The STAT-ON™ Firmware

Pending tasks block

This block is especially useful to avoid timeouts (mainly, during BLE connection or serial communication) while executing actions that may take a significant amount of time. The tasks that may be executed inside this block or function are discussed in the following points:

Erase all data task

When a "deleting all data" request is received from Bluetooth, the pending tasks block waits until the entire Flash memory is completely erased. Therefore, no other task within this block is performed. If any other action is requested on the BLE service, the device notifies that is busy erasing the Flash and discards any request. While this task is being executed, the RGB LED is in blue color (without blinking).

RTC interrupt task

Several actions are systematically done every certain time and for correct synchronization, an RTC is used. This counter is incremented every second and is used for:

- The correct management of the alarms: it checks whether the alarms have been configured via Bluetooth or if any alarm should trigger or is ongoing.

- Battery level checking: Every 10 seconds the battery level is read from the fuel gauge.

- Average current test: Every 10 seconds the current value is read from the fuel gauge. This value is used to know whether the device is in charging state or not.

Additionally, every second the following conditions are checked:

- LED blinking:

  ○ If the device is connected to the Smartphone, the blue LED blinks every second.

  ○ If the device is assessing symptoms, the green LED blinks every second.

  ○ If the battery is beyond 20%, the pink LED blinks every second.

  ○ If the device doesn't have all the configuration parameters, the LED blinks white every second. Otherwise, the LED does not blink. Indicating that the device is in battery-saving mode.

- Serial communication blocked: When movement has been detected and the ST has been powered, the nRF microcontroller expects new data every minute or the indication that it has no more data. If the ST is woken up but there hasn't been any communication for 90 seconds, the ST is switched off. This feature has been implemented to avoid a limbo state.

Data request – BLE task
When the external Smartphone requests new data, the nRF transfers all the data from the Flash memory minute by minute.

Serial communication messages task
This task is responsible for the serial communications in the system. There are communications between the system and the App installed in the Smartphone for configuration purposes and internal communications between both microcontrollers.

- The date and time are sent to the device once the App is connected. Then, the flag indicating that there is a valid timestamp is set. Nevertheless, the timestamp will be sent only when the ST is switched on, and this happens only when movement is detected. It can be updated at any time and can be read by the App via BLE.
- The patient configuration parameters are sent after the date and time message. Some patient-specific parameters are required for the correct running of the algorithms. This information is sent and can be updated and read at any time by the App.
- The communication between both processors is very important and messages via the UART communication channel are shared:
  - **Start recording data:** The nRF microcontroller will transfer all the configuration parameters to the ST to start recording movement data and executing the algorithms. This is also used for checking that both microcontrollers have their timestamp correctly synchronized.
  - **Transfer results:** every minute, the ST will transfer the recoded data and results to the nRF, which is in charge of storing them in the internal memory.
  - Other **data transfers for debugging purposes** (Error logs, time synchronization checking).

Save FLASH information task
It is triggered every time new data has been received from the ST microprocessor and must be stored. This task updates also the Flash memory general information.

Power management block
When executing this block, the nRF microcontroller is entering in low power mode and waits for certain events to wake up. This functionality is provided by the manufacturer.

The possible events for the wake-up condition are:

- BLE events (message received, disconnection, timeout, etc.).
- LIS3DH's interrupt pin for movement detection.
- Button pressed.
- ST pin activation to notify that there's new data.
- ST pin activation to notify that there's no more data to be stored and it is ready to be switched off.
- RTC timer interrupts, occurring every second.

3.4.1.2 The Flash memory

The Flash memory is used to store all the data provided by the ST microprocessor to the nRF microcontroller and, afterward, deliver it to the mobile App. The Flash memory has been mapped in several sectors as follows:

- **Sector 0**: It contains general information (the address of the first empty slot to store new data and the address of the slot that will be delivered to the App when it would request it). Once data is requested, this last address is incremented. It also contains the current patient-specific configuration parameters.
- **Sector 1**: License-related information.
- **Sectors 2–253**: Data. This space is able to store data for nearly 1 year in a continuously saving data regime.
- **Sector 254:** Alarms information.
- **Sector 255**: Sector reserved for error logging and debugging purposes.

Once the nRF is powered up, it initializes the data structure on the Flash memory. This initialization involves:

- Read and check the manufacturer ID, device ID memory interface type, and device ID density.
- Read the first sector to update the patient-related information.

3.4.1.3 The LIS3DH accelerometer

Two specific layers (lis3dh_driver and nrf_LIS3DH) provide the programming and the proper interface with the LIS3DH accelerometer:

The lis3dh_driver
The manufacturer of this accelerometer provides a nearly complete driver. The programmer has to implement two functions: LIS3DH_ReadReg and LIS3DH_WriteReg. These are microcontroller dependent and are implemented in the following nrf_LIS3DH file. The functions allowing the setting of the internal registers ACT_THS and ACT_DUR have been specifically implemented since they were not included in the driver.

The nrf_LIS3DH file
This file provides the initialization, configuration of the LIS3DH, and interface pins. Given that the purpose of this accelerometer is to detect movement and notify the nRF, it is set in low-power mode and low-frequency measurements (10 Hz). Regarding the full scale, it is set at 8G.

Detecting movement is the main function of this accelerometer. To decide if there is movement or not, a threshold of 0.04G has been established with a duration of two consecutive readings. When movement is detected, the INT1 pin is set.

This driver also contains the functions that will be used in the test of the accelerometer, involving a new configuration and reading of new measurements.

3.4.1.4 The management of the ST microprocessor

The nRF microcontroller is in charge of the ST microprocessor management. Given the complexity, the communication between both processors is different from other peripherals. The management is done through two specific functions: ST power management and ST serial communication.

ST power management
The nRF can enable the pin in the LDO that supplies the power to the ST (see Figure 3.3). Since the ST is switched off by default, to reduce power

consumption, it is only awakened whenever the nRF-associated accelerometer detects movement.

Once awaken, it starts the communication between the ST and the nRF. After a while, when the ST sets a certain pin high to indicate that there is no more movement, the nRF cuts the supply of the ST. If by, any chance, the communication gets stuck for over 90 seconds, the ST is switched off.

ST communication management
A module that implements the protocol designed for this application has been developed from scratch. To avoid unnecessary data, the ST will not be awakened until the device has had the timestamp updated at least once and the patient leg length's value is set. In addition, if the device is being synchronized or charged, it will stop assessing symptoms.

3.4.1.5 Alarms
The alarms module allows the user to implement an alarm using three parameters:

- **Mode**: Use the buzzer, the vibrator, or both.
- **Beeps**: Select the number of beeps.
- **Millis**: Duration of the beeps and the silent period in between the beeps.

The alarm configuration by default is Buzzer, two beeps, and a duration of 150 ms. The alarms are triggered once an alarm matches the timestamp and only are executed once.

3.4.1.6 Fuel gauge
To control the battery status, the BQ27441 fuel gauge has been used, and the necessary interface software has been developed (files bq27441_driver and bq27441_fuelgauge). The first contains the functions that deploy the commands, protocol, and timings necessary to interface the BQ27441 and the other one contains the initialization of the BQ27441 and its configuration according to the battery used.

To track the battery level, it is requested by the BQ27441 every 10 seconds. If the battery level has changed, the Smartphone (if it is connected and notifications are enabled) is notified. On the other hand, every 10 seconds the average current is read from the BQ27441. This value is used to be aware that the device is being charged and therefore the ST can be switched off.

3.4.1.7 Bluetooth protocol
This device has four services implemented for the correct management of Bluetooth BLE: the Generic Access Service (generic information about the

62 The STAT-ON™ Industrialization Pathway

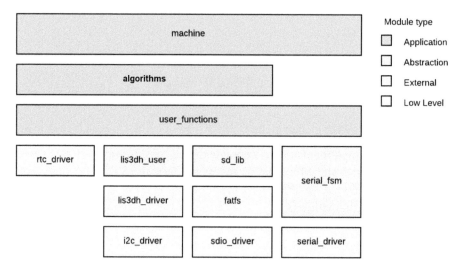

Figure 3.12 ST firmware code architecture.

device), the Generic Attribute Service (defines the GATT hierarchical data structure, the one used on BLE devices), the BAS service (the battery service to monitor its status) and the H4P (holter-for-parkinson) service, designed from scratch to accommodate the nRF to the needs of the new sensor.

3.4.2 Firmware for the STM32F415RGT6

The firmware for the ST microprocessor includes many of the algorithms already developed for the prototype obtained in the REMPARK project and have been refined according to the new structure. So, the main objective was to adapt them to the newly redesigned hardware and allow them to share algorithm results with the rest of the system (nRF microcontroller).

The firmware updating consists of adding a serial communication port to transmit processed data to the nRF and developing a driver for the newly used accelerometer (LISD3H).

3.4.2.1 Firmware code architecture

The ST firmware code architecture is organized into three different layers, as it is presented in Figure 3.12.

- The **application layer** is responsible for the execution of the final user actions. All of the actions executed in this layer shall not depend on peripheral configuration or data adaptation.

- The **abstraction layer** is used to adapt the information from low-level layers to the application layer. In this layer, a specific device setup and data adaptation are performed.
- The **low-level layer** is very dependent on the specific hardware used. Each action relies directly on the different peripherals used.

3.4.2.2 Code modules

A code module is a piece of source code focused on the implementation of certain defined functionalities. The ST firmware includes a number of them. Some modules were designed from scratch, but some others were adapted from the manufacturer's library:

- **Driver_I2C**: it handles the I2C peripheral in the system, allowing access to the accelerometer device lis3dh.
- **Driver_serial**: it provides communication with the nRF microcontroller using the USART peripheral.
- **Serial_fsm**: this module implements the defined communication protocol between ST and nRF processors.
- **lis3dh_user**: this module abstracts the use of the vendor accelerometer library to the application.
- **User_functions**: this is a set of functions used at the application level to handle all the actions that interact with the lower layers. These functions involve, for example, the SD read/write actions, the serial communication, or the accelerometer data acquisition.

3.5 Device Mechanical Design

The enclosure of the product is one of the most important elements and requires careful consideration of its design. Some generally considered characteristics are: it must offer a friendly image, must have a discreet color, and must be usable.

Additionally, it must guarantee robustness against shocks, the enclosure must protect the internal circuitry against dust and water to comply with the IP65 standard. Moreover, as it has been discussed, it must offer solutions to be able to interact with the user by means of a button.

The sensor will have two multicolor LEDs to indicate the device's state. The main specifications defined for the STAT-ON™ case are described in Table 3.4. In these specifications, the characteristics of the

Table 3.4 Electromechanical requirements.

TAG	Component	Requirements definition	Required	Casing design
1	Case	Overall dimension Depth × width × heigh	–	90 × 63 × 21.5 mm
		IP	65	65
		Material	–	ABS
		IK	07	09
2	Battery	Load/charge	1 Ah	
		Overall dimensions	53 × 35 × 5	
3	Push button	Usage		Resistive keypad
6	Buzzer	Overall dimensions (mm)	Ø10 × 3	OK
7	Vibrator	Overall dimensions (mm)	10 × 10 × 12	OK
8	Coil	Overall dimensions (mm)	30 × 30× 1	OK
9	PCB board	Overall dimensions (mm)	73 × 51,25 × 1.65	OK
10	Intermediate ring	Ensure IP	IP 65	
11	Threaded insert	Ensure well joint		

electromechanical components, assembled in the custom plastic casing, are defined.

One of the main items to consider in the mechanical design of the plastic casing is the degree of protection against dust and water intrusion. For STAT-ON™, the level is IP65 (see Table 3.5).

3.5.1 Components selection

According to the already defined requirements, the electromechanical components were defined as part of the redesign process. In concrete, the resistive keypad, the sealing strip, the ironmongery/inserts, and the complete housing were specified.

3.5.1.1 The resistive keypad

For the resistive keypad selection, a test with different users was carried out. Finally, the keypad with 230 µm of the gap was selected. For the adhesive paste, a stronger material has been used to guarantee the tightness of the enclosure according to the IP65 regulation. The membrane is done from polyester (0.15 mm) and the size is 80 × 13 mm^2. Figure 3.13 shows the details and the shape.

3.5 Device Mechanical Design 65

Table 3.5 The IP code for STAT-ON™ is IP65.

IP 65 First digit – Protection against solids	IP Second digit – Protection against water
 A round body, 1.0 mm in diameter, must not be able to penetrate.	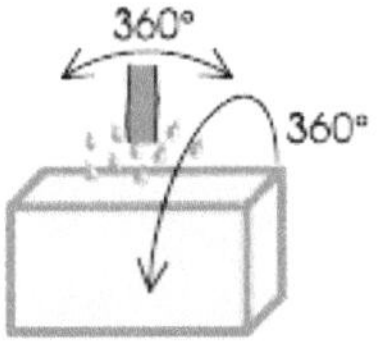 A water jet directed at the enclosure from any direction must not have any harmful effects.

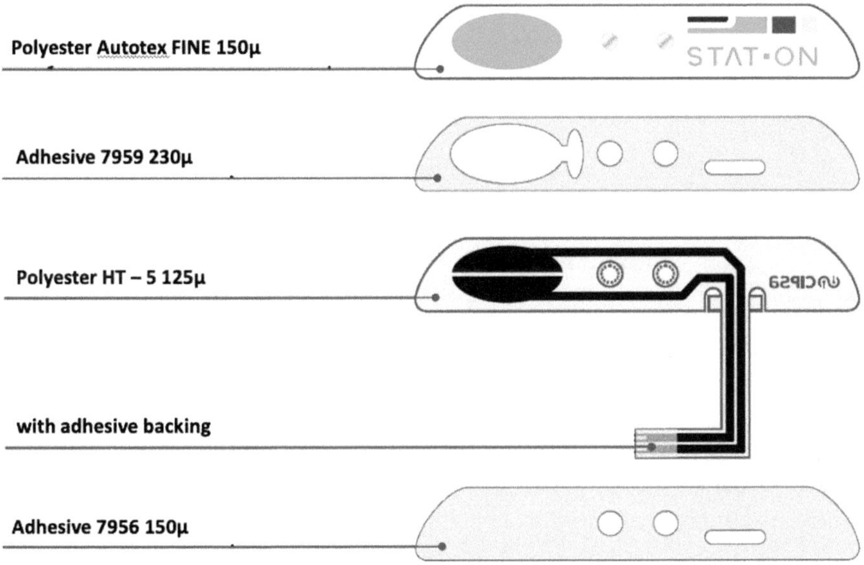

Polyester Autotex FINE 150μ

Adhesive 7959 230μ

Polyester HT – 5 125μ

with adhesive backing

Adhesive 7956 150μ

Figure 3.13 View of resistive keypad.

3.5.1.2 The sealing strip

The sealing strip must guarantee that the device is waterproof. The design of the sealing strip must fit the groove constructed in the enclosure for this purpose.

A total of five materials were tested for massive production. Two of them were made of tough silicone and were directly rejected since the enclosure was deformed when screws were inserted. The remaining three materials were: Bisco BF-1000 (white), Bisco HT-800 (black), and Bisco HT-840 (grey). The last one was also tested with glue on one of the sides for better fixation. Technicians found out that the assembly of the sealing strip with glue was uncomfortable and several times had to remove it from the enclosure since it stuck to the walls of the enclosure.

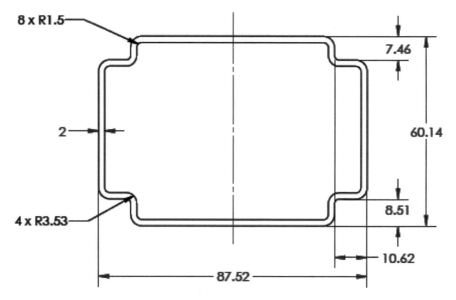

Figure 3.14 Sealing strip shape and dimensions.

Different waterproof tests were performed. Although the three materials worked fine, the most reliable was Bisco HT-840. See Figure 3.14 for shape details.

3.5.1.3 Ironmongery – Inserts
A heat installation insert was selected due to the fact that it ensures a longer useful life than the press insertion models. The selected model is used to increase resistance to torque. See Figure 3.15 for insert performance details and Table 3.6 for the details on inserts.

3.5.1.4 Housing design
With all the selected components, the housing was designed, consisting of two parts. See the initial design idea in Figure 3.16.

The main component of the enclosure is the thermoplastic polymer, called Acrylonitrile Butadiene Styrene (ABS) since it has good mechanical and impact strength, combined with ease of processing (reducing the injection costs).

This material is operable with a wide range of temperatures (-20 °C until 80 °C), more than enough for the purpose of this case. The wall thickness that is possible to obtain, depending on the manufacturing process finally decided, has been taken into consideration (> 1 mm if a silicone mold is used

3.5 Device Mechanical Design   67

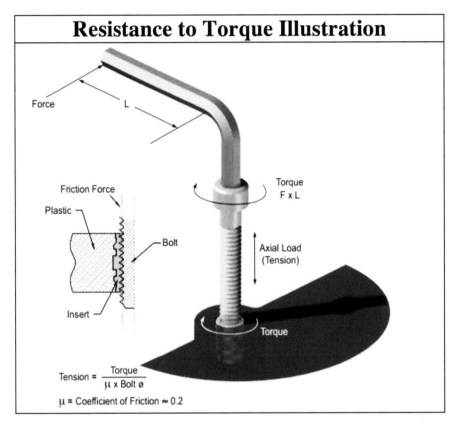

Figure 3.15 Insert performance.

Table 3.6 Inserts parameters.

INSERTS Heat Installation	
Manufacturer/Supplier	Spirol
Model	INS 29/M2,5
Reference	150924
Metric size (mm)	2.5
A (mm)	4.7
P (mm)	3.9
Recommended hole (mm)	4
L (mm)	3.5

68   The STAT-ON™ Industrialization Pathway

Figure 3.16   Designed housing. Initial design.

or between 1.143 and 3.556 mm if we decide on the option of injected ABS).

The final industrial design of the housing is shown in Figures 3.17 and 3.18. The surface of the device enclosure was designed, in one of its first prototypes, totally smooth. However, it has been proved that this surface type caused scratches very easily. In addition, the continuous use of the device fouled the surface in contrast to the white color resulting in a degraded image of the sensor. For that reason, a matt surface and a darker white color were finally decided.

3.5.2 Enclosure industrialization

In the previous section, the complete design of the STAT-ON™ enclosure has been presented, and is ready for industrial production. The plan was to produce several series in an aluminum mold.

Adjusting of the thickness of the plastic walls is very important to avoid problems coming from the extraction of the enclosure from the mold and the possible sudden and aggressive changes of temperatures that could severely affect the shape of the case.

Figure 3.19 shows the final enclosure, ready for its industrialization, with the final measurements. It is interesting to note that the final shape of the enclosure was maintained after the satisfactory results obtained in a usability test.

The thickness of the wall must be as uniform as possible. The part of the screws had a solid area of plastic that had to be redesigned from the original design, to remove the plastic without affecting the structure of the enclosure. Figure 3.20 shows the details of the parts allocating the screws.

3.5 Device Mechanical Design

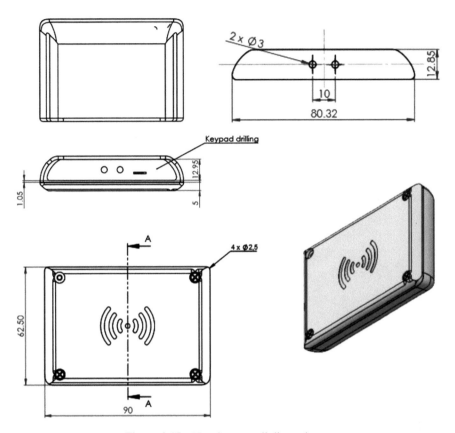

Figure 3.17 Housing overall dimensions.

It has also been selected key points to eject the plastic of the mold with the aim of not generating irregularities on the shape of the case in the ejection phase. Figure 3.21 shows the points of ejection of the case.

Figure 3.22 shows the final industrialized enclosure for STAT-ON™ commercialized device.

3.5.3 The belt

A belt is necessary to fix the STAT-ON™ in its correct position, in the waist, and slightly displaced to the left. The belt must have a pocket for an easy insertion of the device giving access to the top bouton for its pressing, when necessary, with a velcro tap for easy fixation.

The belt is made of Polyester (94%) and elastane (6%). Its fabric allows a complete adjustment to the body while being comfortable. A hook and loop

70 *The STAT-ON™ Industrialization Pathway*

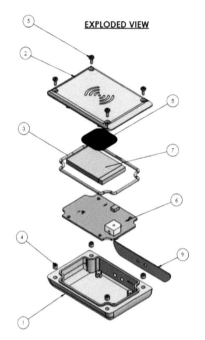

NO. OF ELEMENTO	NAME OF ELEMENT	QUANTITY
1	PARK-IT_CASE_BASE-PART	1
2	PARK-IT_CASE_TOP-PART	1
3	PARK-IT_HOUSING_O-RING	1
4	PARK-IT_HOUSING_INSERT-	4
5	STAINLESS STEEL SCREW M2,5 X 6	4
6	PARK-IT_PCB	1
7	PARK-IT_HOUSING_BATTERY	1
8	Park-It_Coil	1
9	PARK-IT_HOUSING_RESISTIVE-KEYPAD	1

Figure 3.18 View of the different parts forming the complete housing.

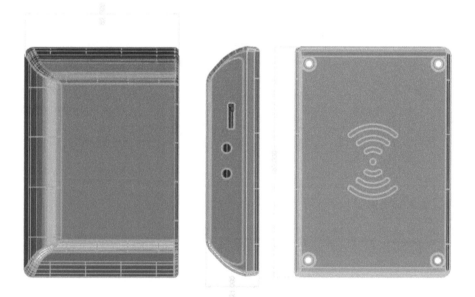

Figure 3.19 General measurements of the box.

3.5 Device Mechanical Design 71

Figure 3.20 Bottom view of the enclosure showing the part of the screws.

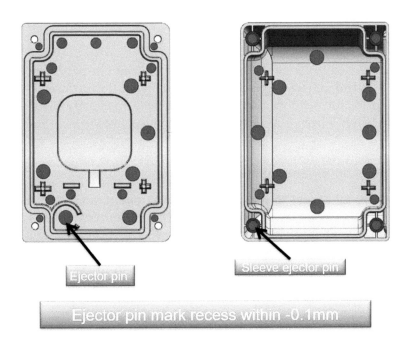

Figure 3.21 Ejection points.

Figure 3.22   Final industrialized enclosure of the STAT-ON™ device.

fastener is used to fasten the belt securely. The belt has passed the Oeko-Tex® Standard 100 tests, guaranteeing the safety of the textile and the perfect compatibility with the skin. The belt can be worn directly on the skin or above a t-shirt. Figure 3.23 shows the final industrialized belt.

3.5.4 Packaging and labeling

To commercialize and ship correctly the STAT-ON™, a specific package, containing the sensor, has been designed. The accompanying belt is packaged in a specific separated bag, since in this way, is easier to serve a sensor accompanied by several belts to a given customer.

The package has been specifically designed to fit the sensor and be optimal with the space for shipping. The design consists of three pieces (Figure 3.24):

- Base: one model open size 25.5 × 32 cm in Inverkote Mat paper of 350 g in 4+0 inks. Matte laminate on one side. Self-assembling die cut.

3.5 Device Mechanical Design 73

Figure 3.23 The belt.

- Band: one model open size 29.5 × 12.4 cm in Inverkote Mat paper of 300 g in 4+0 inks. Matte laminate on one side. Split + apply adhesive tape to one end and close.
- Nest: one model open size 19 × 16.8 cm in Inverkote Mat paper of 350 g in 4+0 inks. Matte laminate on one side. Die-cut with hole for sensor (to be inclined) and adhesive inside the base.

Figure 3.25 shows the aspect of the assembled packaging, ready for distribution. The package's total weight is 330 g and it has an eco-solvent print with matte polypropylene.

The sensor labeling must be according to the medical device regulation and contain the required data. Figure 3.26 shows the label for the STAT-ON™ in the left side and the device with its label on the right.

3.5.5 Battery charging system

The charge of the STAT-ON™ battery is done wirelessly and for this operation, a standard wireless base charger was selected, with the following main features (Figure 3.27):

- Qi-certified charging pad,
- stylish and portable design
- compatible with any Qi-enabled smartphone or device
- including Micro-USB to USB-A cable, and
- requires 2A.

The accompanying AC charger is the GSM12E05-USB, a medical AC charger that is compatible with any charger base with the presented features.

74 The STAT-ON™ Industrialization Pathway

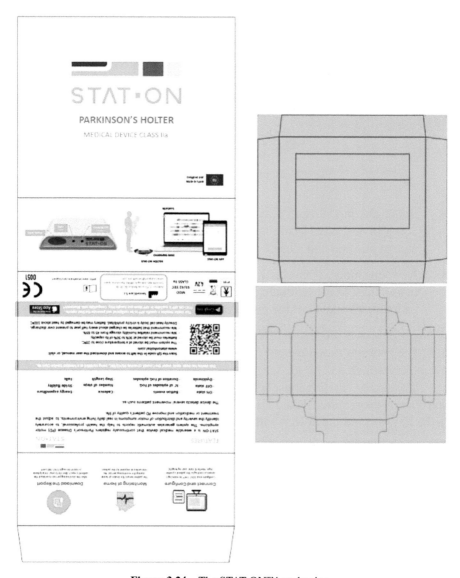

Figure 3.24 The STAT-ON™ packaging.

The AC charger can be used at home since it is compliant with the EN/EN60601-1/ EN/EN60601-1-11.

3.6 Certification and Characteristics

The industrialized and produced final device is presented with the features specified in Table 3.7, and is certified under the standards indicated in Table 3.8:

3.6 Certification and Characteristics 75

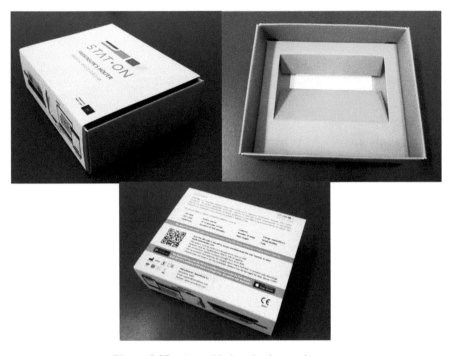

Figure 3.25 Assembled packaging ready to go.

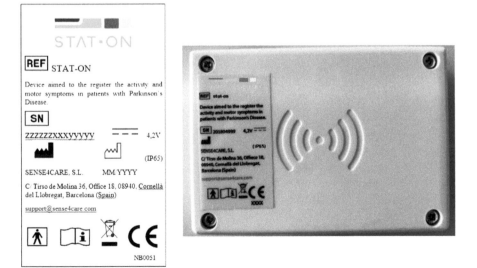

Figure 3.26 The STAT-ON™ labeling.

Figure 3.27 Charging pad aspect and dimensions.

3.7 Conclusions

The present chapter presented the complete redesign and industrialization process of the prototype developed in the frame of the REMPARK project [1], enabling the embedding of the complete developed algorithmic set, based on machine learning techniques.

The achieved product is STAT-ON™, a medical device Class IIa, able to act as a Holter for Parkinson's Disease, detecting and measuring, in real time, the motor symptoms associated with PD [5,6].

The redesign process of the device was based on the knowledge achieved on the REMPARK's prototype, but also considering that the device has to be assembled by a third party and must be manufactured in a series of hundreds of units when commercialized.

The cost of the manufacturing process and compliance with the regulatory standards are crucial challenges. In concrete, the mentioned PARK-IT2 project had a whole work package focused on industrialization, covering the certification and the redesign process of the device.

Very demanding specifications were raised from the redesign of the hardware to the mechanical design with the objective that the REMPARK prototype could reach the market. The change in data flow and the new user interface were major technical challenges. Throughout the process, efforts were made to follow the highest quality standards, and multiple tests were carried out, which in many cases forced long redesign processes. For these reasons, the industrialization stage of the sensor lasted for a long time of work in the different lines of the design. But all this work was worth it and STAT-ON™ was born, a sensor that, from a technical point of view, represented a great advance compared to previous models and the competitive landscape surrounding STAT-ON™, both in performance and in reliability and hardness.

Table 3.7 STAT-ON™ characteristics.

Communications	
Bluetooth specification	Bluetooth 4.0 (Bluetooth Low Energy)
Bluetooth bandwidth	2,4 GHz
Wireless charging standard	WPC v1.1 Qi Industry Standard
Wireless charging bandwidth	100-205 kHz
Electrical features	
Power supply (charger)	100-240 Vac, 0.3-0.6 A, 50-60 Hz
Battery: type	Lithium polymer
Battery: capacity	1100 mAh
Battery: charging time	<6 h
Battery: maximum charging current	500 mA
Battery: maximum discharge current (peak)	135 mA
Average consumption (normal use)	2.5 mA
Physical features	
Height	62,5 mm
Width	90 mm
Depth	21,20 mm
Weight	86 g
Enclosure material	ABS-FR(17) UL94, UV Protection White - Matte
Environment specifications	
Temperature operation range	From 0°C to 40°C
Temperature in charging conditions	From 0°C to 40°C
Storing conditions	The system must be stored at a temperature close to 20ºC and with batteries charged about 30% to 50% of capacity. We recommend relative humidity storage from 45 to 85%. We recommend that batteries be charged about every half year to prevent over discharge. Directly heat cell body is strictly prohibited. Battery may be damaged by heat above 100ºC.
Atmospheric pressure conditions	700 hPa to 1060 hPa
Certification	
Protection against and dust and water	IP65
Battery in medical use	IEC62133
Design, fabrication, and commercialization of industrial electronic controls.	ISO 9001:2015
Medical quality management system and medical devices sales, development, manufacturing, delivery and maintenance including related services	ISO 13485:2016
Medical device certification	CE Marked number: 0051

Table 3.8 Standards affecting STAT-ON™.

#	Standard	Harmonized	Application
1	EN 1041:2008 Information supplied by the manufacturer of medical devices	YES	Used to establish the information needed for product use and general aspects of the presentation of information
2	EN 15223-1:2016 Symbols for use in the labelling of medical devices	YES	Used to set the appearance of graphical symbols included in the labelling of our product.
3	EN ISO 60601-1:2006/A1:2013 Medical electrical equipment. Part 1: General requirements for basic safety and essential performance.	YES	Used for establishing the basic safety and essential performance. Date of cessation of conformity for previous ed. 31.12.2017
4	EN ISO 60601-1-2:2015 Medical electrical equipment. Parts 1–2: General requirements for basic safety and essential performance. Collateral standard: Electromagnetic compatibility. Requirements and tests.	YES	Used for establishing the safety and functionality EMC requirements
5	EN 60601-1-6:2010 Medical electrical equipment. Parts 1–6: General requirements for basic safety and essential performance. Collateral standard: Usability	YES	Used for establishing usability requirements for medical electrical equipment
6	EN 60601-1-11:2010 Medical electrical equipment. Parts 1–11: General requirements for basic safety and essential performance. Collateral Standard: Requirements for medical electrical equipment and medical electrical systems used in the home healthcare environment	NO	Se utiliza para establecer los requisitos y las pruebas para el dispositivo como equipo eléctrico médico y sistemas médicos eléctricos utilizados en el entorno de atención médica domiciliaria.
7	EN 62304:2006+/AC:2008 Medical device software. Software life cycle processes.	YES	Used for establishing the life-cycle of software
8	EN ISO 14971:2012 Medical devices. Application of risk management to medical devices	YES	Used for establishing the risk management process for the product
9	EN 80002-1:2009 Medical devices software. Guidance on application of ISO 14971 to medical device software	NO	Used for establishing the risk management process for the software

3.7 Conclusions   79

#	Standard	Applicable	Notes
10	EN ISO 62366:2008 Medical devices. Application of usability engineering to medical devices	YES	Used to minimize use-errors
11	MEDDEV 2.7.1 (2016) Clinical Evaluation Clinical Evaluation – A guide for manufacturers and notified bodies	NO	Guidance for device Clinical Evaluation
12	EN ISO 14155:2011 Clinical investigation of medical devices for human subjects. General requirements	YES	Applies only Chapter 4 and recommendations for the review of data and medical and scientific information published/available as Annex A.
13	EN 62353:2014 Medical electrical equipment – Recurrent test and test after repair of medical electrical equipment.	NO	Used for establishing the test after repair and preventive maintenance plans
14	RED 2014/53/EU The Radio Equipment Directive	YES	Used for establishing the radio Equipment requirements
15	ETSI EN 300 328 V2.1.1 Harmonized Standard covering the essential requirements of article 3.2 of Directive 2014/53/EU	YES	Wide Band Data Transmission equipment standard.
16	ETSI EN 301 489-1 V2.2.0 Article 3.1b Directive 2014/53/EU - RED	NO	ElectroMagnetic Compatibility (EMC) standard for radio equipment and services; Part 1: Common technical requirements;
17	ETSI EN 301 489-3 V2.1.1 Article 3.1b Directive 2014/53/EU - RED	NO	ElectroMagnetic Compatibility (EMC) standard for radio equipment and services; Part 3: Specific conditions for Short-Range Devices (SRD)
18	ETSI EN 301 489-17 V3.2.0 Article 3.1b Directive 2014/53/EU - RED	NO	ElectroMagnetic Compatibility (EMC) standard for radio equipment and services; Part 17: Specific conditions for Broadband Data Transmission Systems;
19	ETSI EN 303 417 V1.1.1 Wireless power transmission systems Harmonized Standard covering the essential requirements of article 3.2 of Directive 2014/53/EU	NO	Wireless power transmission systems, using technologies other than radio frequency beam, in the 19–21 kHz, 59–61 kHz, 79–90 kHz, 100–300 kHz, 6765–6795 kHz ranges;
20	EN 60601-1-11:2015 Medical electrical equipment. Parts 1–11: General requirements for basic safety and essential performance. Collateral Standard: Requirements for medical electrical equipment and medical electrical systems used in the home healthcare environment	NO	It is used to establish the requirements and tests for the device such as medical electrical equipment and electrical medical systems used in home environments.

References

[1] REMPARK–Personal Health Device for the Remote and Autonomous Management of Parkinson's Disease. FP7-ICT-2011-7-287677. 2011-2014. n.d.

[2] Cabestany J, Bayés À. Parkinson's Disease Management through ICT: The REMPARK Approach. River Publishers; 2017. https://doi.org/10.13052/rp-9788793519459.

[3] Semiconductors N. nRF51822. Bluetooth Low Energy and 2.4 GHz SoC 2021. https://www.nordicsemi.com/products/nrf51822 (accessed September 3, 2021).

[4] ST Microelectronics, Inc ©. STM32F415xx,STM32417xx Data Sheet 2016.

[5] Rodríguez-Martín D et al. A Wearable Inertial Measurement Unit for Long-Term Monitoring in the Dependency Care Area. *Sensors* 2013;13:14079–104. https://doi.org/10.3390/s131014079.

[6] Rodríguez-Martín D et al. A Waist-Worn Inertial Measurement Unit for Long-Term Monitoring of Parkinson's Disease Patients. *Sensors* 2017;17:827. https://doi.org/10.3390/s17040827.

4

The EU Medical Device Regulatory Process: The STAT-ON™

Daniel Rodríguez-Martín and Martí Pie

Sense4Care S.L. – Cornellà de Llobregat, Spain

Email: (daniel.rodriguez) (marti.pie)@sense4care.com

Abstract

This chapter describes the complete regulatory process followed by STAT-ON™ product to comply with the actual European regulation. A complete idea of the steps to be followed, the associated estimated timing, and the material to be considered and prepared are presented. A complete discussion about the Quality Management System of the manufacturing company is also discussed.

4.1 Introduction

The regulatory process is one of the most important challenges that a manufacturer of medical devices must face. This process, which is costly and long, is essential and mandatory for maintaining the safety of patients whatsoever the field is being treated. In this chapter, we treat the main points and the pathway for achieving a medical device certificate for the European market based on the experience obtained in STAT-ON™.

4.1.1 Definition of a medical device

According to the Official Journal of the European Union (OJEU), where the regulations about medical devices are published, a medical device is defined as:

> "…any instrument, apparatus, appliance, software, implant, reagent, material or other article intended by the manufacturer

to be used, alone or in combination, for human beings for one or more of the following specific medical purposes:

- *diagnosis, prevention, monitoring, prediction, prognosis, treatment or alleviation of disease,*

- *diagnosis, monitoring, treatment, alleviation of, or compensation for, an injury or disability,*

- *investigation, replacement or modification of the anatomy or of a physiological or pathological process or State,*

- *providing information by means of in vitro examination of specimens derived from the human body, including organ, blood and tissue donations,*

and which does not achieve its principal intended action by pharmacological, immunological or metabolic means, in or on the human body, but which may be assisted in its function by such means."

Given the importance of this statement, medical devices need to be controlled and regulated by strict rules in order to provide rigorous measurements that are applied in the field of health for several purposes.

Medical devices can be classified depending on the risk concerning the patient in several classes: class I (low risk), class IIa (medium risk), class IIb (medium/high risk), and ending with class III (high risk). Usually, the specific classification of a medical device must be done by an externally certified notified body, except for class I devices.

4.1.2 Directive MDD93/42 and the regulation MDR2017/745

In the recent past, since 1993, medical devices have been regulated under the Council Directive 93/42/EEC of June 14, 1993 (MDD93/42). This directive intended to harmonize the laws relating to medical devices within the European Union. However, medical devices have changed and progressed significantly, and several manufacturers demanded new regulations, notified bodies, and users of medical devices.

Since 2017, the new regulation MDR 2017/745 [1] on medical devices has prevailed in Europe, which derogates the MDD93/42/EEC. At the same time, the regulation set a 3-year moratorium after the date of entry into force, during which those devices certified under the previous regulation continued to be valid. Due to the pandemic, this moratorium was extended until May 2021, the date from which a mandatory adaptation to the new regulatory framework is required.

According to the Legislative Act, the new regulation (MDR2017/745):

> "...aims to ensure the smooth functioning of the internal market as regards medical devices, taking as a base a high level of protection of health for patients and users, and taking into account the small- and medium-sized enterprises that are active in this sector. At the same time, this Regulation sets high standards of quality and safety for medical devices in order to meet common safety concerns as regards such products. Both objectives are being pursued simultaneously and are inseparably linked whilst one not being secondary to the other. As regards Article 114 of the Treaty on the Functioning of the European Union (TFEU), this Regulation harmonises the rules for the placing on the market and putting into service of medical devices and their accessories on the Union market thus allowing them to benefit from the principle of free movement of goods."

In other words, the new regulation improves the EU market functioning, ensuring safety for users and patients, and sets standards for the new quality management system, which empowers the companies to work with their products in the market.

Due to the moratorium application, Europe is living in a curious and exceptional situation since all manufacturers must adapt their medical devices. This has complicated the processes and logistics of the European Notified Bodies and manufacturers to accommodate their devices to the MDR 2017/745.

This, along with the Brexit situation, has made the manufacturers to also adapt their medical devices to the British authorities' requirements for obtaining the new UKCA certificate. This scenario has provoked a serious saturation in the notified bodies' activity, prolonging the periods to achieve a medical device certificate. Due to this situation, the period to certify a medical device can easily be 1 to 2 years.

It must be noted that STAT-ON™ is a medical device, class IIa, certified according to directive 93/42/EEC, that requires an adaptation to the new regulation. This process must be done before May 2024 to comply with the MDR 2017/745.

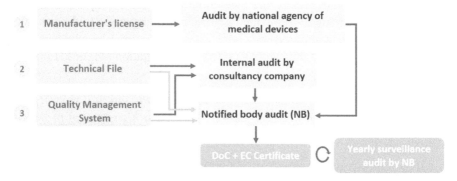

Figure 4.1 Main processes and steps for achieving the EC certificate.

4.1.3 The regulation processes

The regulation processes involved in the regulatory frame (for both the directive 93/42/EEC and the MDR 2017/745) are, in essence, very similar and are described hereafter. It must be noted that these processes affect the STAT-ON™ certification.

In order to achieve the EC Certificate and the Declaration of Conformity of the product, there are three main steps that any manufacturer has to follow:

- the manufacturer's license.
- the technical documentation.
- the quality management system.

The manufacturer's license is provided by the National Agency of Medicines and Medical Devices, after an audit that includes the procedures, the manufacturing process, and the people that are in charge of each process. This manufacturer's license is essential given that it enables the manufacturer to manufacture the medical device through an official agency and is also mandatory for the documentary audit to be done by the notified body.

The technical documentation and the quality management system are documents that gather all the necessary references for the preparatory internal audit permitting the audit done by the notified body. All these documents refer to the product and the company that manufactures and places the device in the market. A scheme of the processes and steps is shown in Figure 4.1.

The generation of all the mentioned documentation can take a long time (sometimes from 6 to 9 months), and the audit for the manufacturer's license done by the national agency of medical devices can take time as well (around

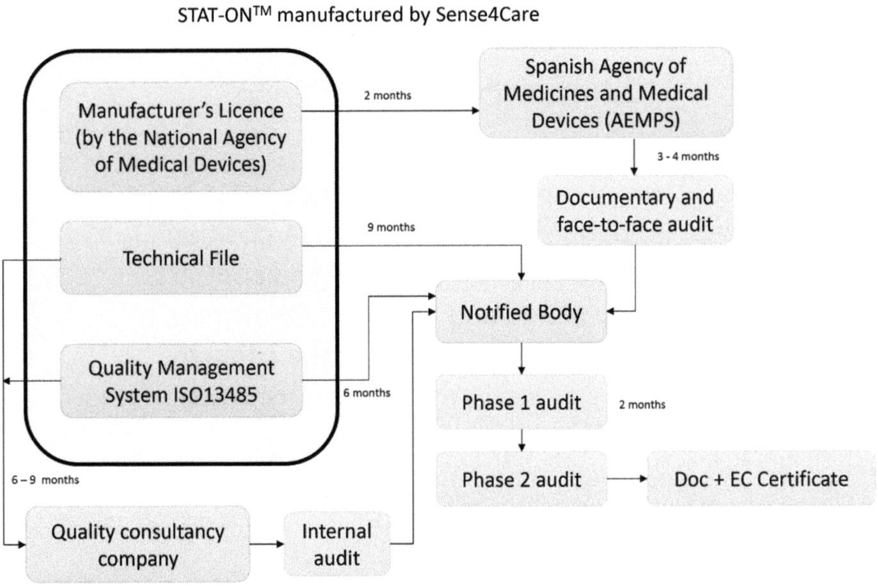

4.1 Introduction

Figure 4.2 Whole detailed diagram process for achieving the EC Certificate.

two months since this is requested could be a good estimation). The audit to be done by the notified body can take extra time (one year is a good estimation, given that there are many documents to prepare). Finally, a review and a face-to-face audit process are required. These estimated times can be extended given the crowded scenario that notified bodies are facing. The European Commission has considered extending the adaptation of the medical devices from May 2024 to the year 2028 with conditions for manufacturers such as a rigorous commitment to adapt the medical device to the MDR.

In the case of the STAT-ON™, two days and a half were needed for the notified body audit. Before this process, a regulatory expert company also performed an internal audit of the manufacturer and the company that places the device in the market. It must be considered that the manufacturer can subcontract a specialized company in soldering, board production, and assembly procedures, which is the case of STAT-ON™. This company is also under strict rules and quality requirements. In this specific case, the company must comply with ISO13485 for manufacturing medical devices.

The complete process is depicted in Figure 4.2, where the presented timing is estimated. The time and delay strongly depend on the notified body's activity saturation and the manufacturer's ability to generate the complete requested documentation.

Manufacturer's License

- License request to the Agency
- Personnel documentation (CV, position...)
- Subcontractor agreements and contracts
- Facility plants and additional information
- Normalised procedures of the company
- Activity procedures, controls, reviews...
- Environmental conditions
- Manufacturing procedures
- Personnel responsibilities
- Company organigram

Figure 4.3 Manufacturer's license documentation.

Each one of the procedures is described in the next sections. In Section 2, it is summarized the manufacturer's license process. In Section 3, it is briefly described the content of the technical documentation, and in Section 4, the quality management system is presented.

4.2 The Manufacturer's License

The manufacturer's license is the necessary first step in the certification process that enables a manufacturer to place a medical device in the market. The manufacture's license is provided by each country's national agency of medicines and medical devices. In the case of Spain, it is driven by the Spanish Agency of Medicines and Medical Devices (AEMPS).

The complete list of the documentation to be prepared by the manufacturer company is in Figure 4.3, and some additional details are given in the following text. The requested documents are some of the documents that must be included in the quality management system (QMS) (see Section 4). Thus, when the QMS is performed, the manufacturer's license documents are implicit within the QMS.

The preliminary action to be performed is to request the national agency to start the manufacturer's license process, and it is necessary to elaborate and prepare a list of documents that are mainly related to the description of the manufacturer company: the description of the involved personnel, details about the contracts and agreements with subcontractors, information about the plants, the description of the procedures and controls in the company, details on the manufacturing processes, the complete organization and organigram of the company, etc.

The national agency (the AEMPS in the case of Spain) performs a documentary audit of these documents and also a face-to-face audit in the facilities of the manufacturer or in the facilities of the manufacturing process.

The manufacturer's license is valid for 5 years and must be renewed and audited once the expiration date arrives. Usually, the face-to-face audit takes no more than one day. It consists of an initial meeting, a visit to the subcontractor facilities (if any), a documentary review (related to the manufacturer and the subcontractor) of aspects related to the subcontractor, and the final reading and signing of the inspection report.

During the meeting, many points and aspects are checked and reviewed. Among them:

- Some specific aspects of the software (installation, validation protocol, etc.).
- Surveillance system, notification, and evaluation of adverse events.
- Work instructions in case of incidents and nonconformities treatment procedures.
- Procedures in case of a market product withdrawal.
- Company organigram and technical manager responsibilities.
- Manufacturing procedure, installation, and maintenance. Risk analysis management report.
- Design and control change procedure.
- Identification, traceability, and inspection state of the products procedure.
- Documents file system procedure (contract with subcontractors, fabrication order, product label model, providers follow-up, etc.)

An inspection is also performed on the subcontractor to check if they comply with ISO13485 (for manufacturing medical devices). However, this inspection is not mandatory if the subcontractor owns the ISO 13485 certification.

If this inspection must be done, all the manufacturing processes are inspected and all the documents assigned to each manufacturing machinery are checked, as the ISO13485 indicates. The auditor can request additional documents from the subcontractor (packaging methods and instructions, calibration of the equipment, calibration plan, calibration certificates, machines' maintenance, etc.)

The estimated required timing is indicated in Figure 4.2; the generation of all the documents and the final audit can take from 8 to 12 months (6–9 months to generate the documents and 2–3 months to perform the audit process). If the procedure is successful, the certification, with a validity of five years, is provided to the company at the end of the audit. This certification is mandatory in the final audit performed by the notified body.

4.3 The Technical Documentation

The technical documentation is an important part of the regulatory process. In essence, it is a complete set of technical information about the medical device, comprising:

- The specification of its components: all the schemes, graphics, manuals, industrial design plans, codes, etc.

- The laboratory tests performed and succeeded, including their certification.

The technical documentation also includes the risk management file, the labeling, the required user manual of the device, and the clinical evaluation, including usability, endorsement by experts, and state-of-the-art. Finally, a commercial part for surveillance of the device after commercializing is also incorporated, along with the declaration of conformity and the final certification.

This block of documents is very extensive but provides all the details of the device aligned with the main technical standard IEC60601-1 for medical devices. The technical documentation structure can be divided into three main blocks: summary, detailed documentation, device modifications, and follow-up (see Figure 4.4).

4.3.1 Part A: The summary

Part A of the technical documentation contains the main information of the medical device, synthesized in a single document for a better and quicker

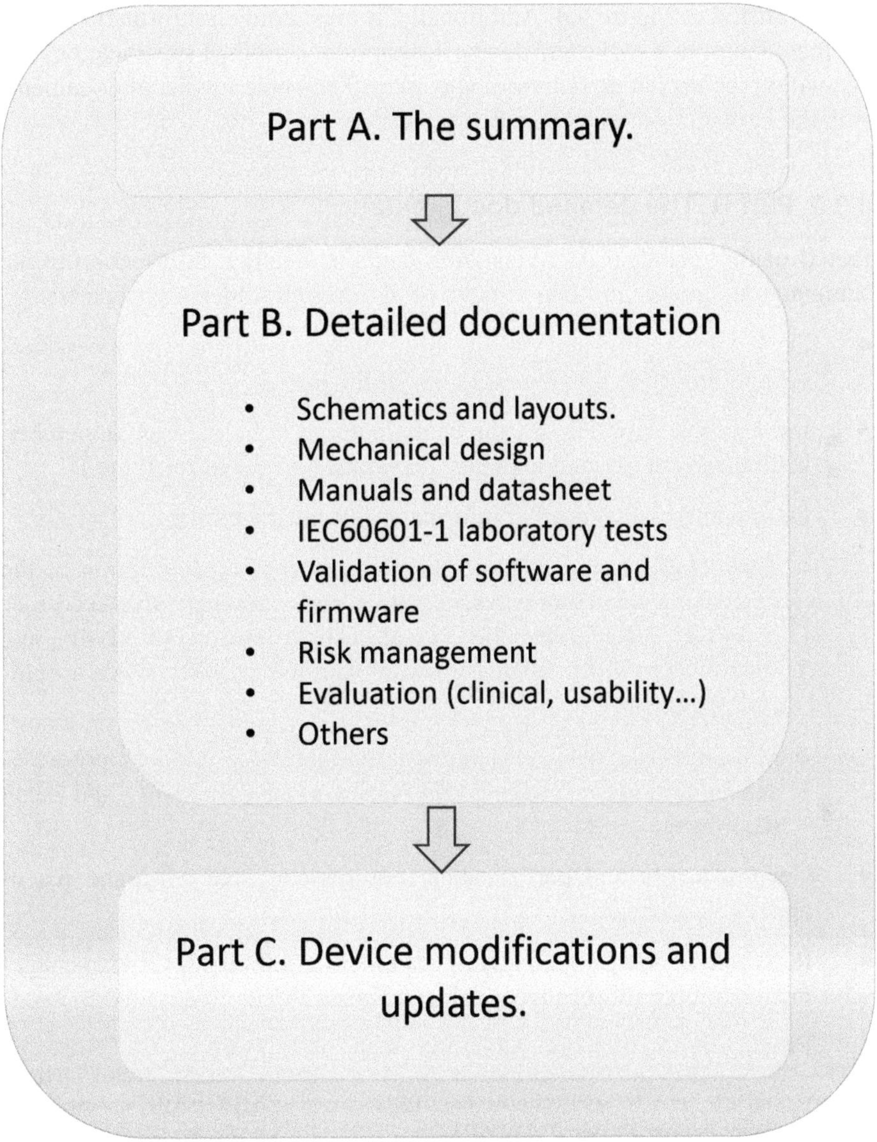

Part A. The summary.

Part B. Detailed documentation

- Schematics and layouts.
- Mechanical design
- Manuals and datasheet
- IEC60601-1 laboratory tests
- Validation of software and firmware
- Risk management
- Evaluation (clinical, usability...)
- Others

Part C. Device modifications and updates.

Figure 4.4 Structure of the technical documentation.

understanding (its classification, the purpose of use, contact information, the brief information of the manufacturing process and subcontractors, and how the company will deal with postmarketing surveillance).

This document is a guide for the auditors and is crucial for the manufacturer. The document is structured according to the detailed documentation (part B) listed in Figure 4.4. Additionally, it must contain information about the manufacturer, a scope and device description, a product specification, and an analysis of similar devices on the market. The complete list of documents is in the next section when Part B is presented.

4.3.2 Part B: The detailed documentation

Part B of the technical documentation contains detailed information in the summary document. Part B is structured in different folders or subparts:

- Device description and specification, including variants and accessories with reference to previous versions of the device.
- Information supplied by the manufacturer (labeling, serial number, instructions of use and manuals, declaration of conformity, etc.)
- Design and manufacturing information. It must include:
  - The device design and specifications (technical description of the device and accessories, schematics and drawings, the necessary materials, calculations and critical design elements, the design and specification of the related software, and the finished device specifications, etc.).
  - Manufacturing details (manufacturing facilities, suppliers, processes and conditions, packaging and sterilization, traceability and batch records, etc.).
- General safety and performance requirements (checklist and list of applicable standards).
- Risk/benefit analysis and risk management (methodology, risk management summary of results, and final statement).
- Product verification and validation, with safety tests:
  - Safety of Electromedical Equipment – Tests performed and summary report as per EN 60601-1.
  - Electromagnetic Compatibility – Tests performed and summary report following EN 60601-1-2.
  - Biocompatibility of applied parts.
  - Functionality and efficacy tests.
  - Device lifetime. Stability/aging tests.

- ○ Usability – tests performed and summary report following EN 62366.
- Clinical data. It must include a clinical evaluation report with conclusions concerning risk/benefit and a plan for postmarket clinical follow-up (PMCFU).
- Additional information must be provided in some specific cases:
  - ○ Devices containing medicinal substances.
  - ○ Devices or derivatives manufactured utilizing tissues or cells of human or animal origin.
  - ○ Devices are composed of substances that are absorbed or locally dispersed in the human body.
  - ○ Devices placed on the market are sterile or in a defined microbiological condition.
  - ○ Devices with a measuring function.
  - ○ Devices are to be connected to other devices in order to operate as intended.
- Final conclusion and Declaration of Conformity, with the EC Conformity Evaluation Procedure (done by the notified body).

The following section presents the concrete case of the STAT-ON™ medical device, and some related documents in its technical documentation are described.

4.3.2.1 STAT-ON™. The device description and specifications

This set of documents gathers the main description of the device, explaining the different parts and briefly describing how the system works. The mechanical and electrical diagrams accompany this documentation. The mechanical diagrams show the plans of the enclosure, the sealing strip, and the button membrane of the device. The electrical diagrams are composed of the schematics and layout of the electronic circuits. It is also necessary to include all the billing of the materials, their provider, and their cost.

The mechanical enclosure details have already been presented and discussed in Chapter 3, and the different figures included there show the enclosure design, its dimensions, the different parts, and the final aspect of the commercial device.

Furthermore, the document informs about the classification of the medical device, compliance with 93/42/EEC, and the category of the device.

> The product pertains to the following devices' category: **04 Electromedical/mechanical.**
> The product pertains to the following subcategories: **05 MD1301 Monitoring devices of nonvital physiological parameters.**

It also reported data about the notified body:

IMQ ISTITUTO ITALIANO DEL MARCHIO DI QUALITÀ S.P.A. – NB 0051
Via Quintiliano, 43 20138 – Milano. Italy
Tel: +39 02 50731
Email: info@imq.it

The complete electrical diagrams are also part of the technical documentation. A complete description of the hardware electronics is contained in Section 3 of Chapter 3, where the final architecture and details of the redesign are conveniently detailed. Figure 4.5 shows, as an example, the schematics corresponding to the related Nordic nRF51822 processor circuitry.

4.3.2.2 STAT-ON™. The information supplied by the manufacturer

This document includes all the information provided by the manufacturer (Sense4Care SL, in this case) to the customer. This documentation must also include the labeling (per the regulation) and the user manual.

Fulfillment of labeling requirements

The content of product labeling (label unit, packaging labels, and instructions for the use) has been established per regulatory requirements of directive 93/42/EEC, Annex I.13 and with the requirements of the EN 15223-1:2016, EN 1041:2008 (it is necessary to remember that STAT-ON™ is a medical device under the directive 93/42/EEC, in adaptation process to the MDR 2017/745).

Product markings: Serial number label

Figure 4.6 shows the produced labeling for STAT-ON™, which contains the product lot number and some graphic symbols (following standard EN 15223-1). The serial number can have several combinations, but the lot and the serial number are mandatory. In the STAT-ON™ case, it comprises three-part codes: month and year of fabrication, lot number, and the serial number (SN). At its turn, the SN is composed of the fabrication date (zzzzz), the lot

4.3 The Technical Documentation 93

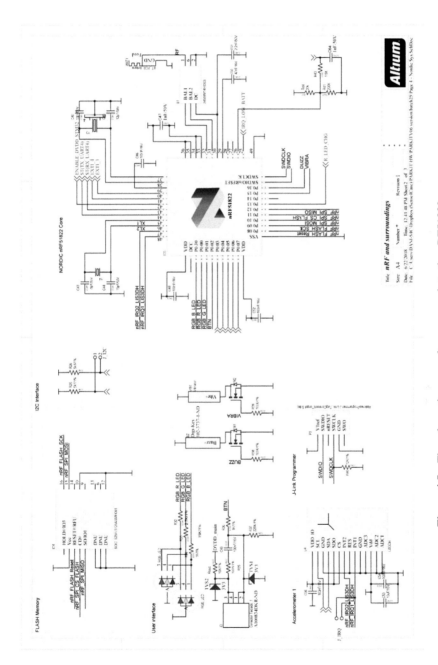

Figure 4.5 The schematics example corresponds to the nRF51822 processor.

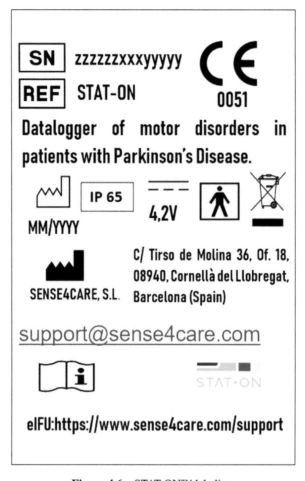

Figure 4.6 STAT-ON™ labeling

number (xxx), and the order number of the concrete device in the lot (yyyyy). Some more details are in Section 5.4 of Chapter 3.

Declaration of conformity
The declaration of conformity is an official document approved by the notified body, which is sent to the customer to confirm the compliance of the device with the set of rules and standards declared in the document. The document has to be dated, signed by the main responsible for the company, and accepted and validated by the notified body.

In concrete, this document establishes the following items:

- Description of the product family, indications, and intended use.

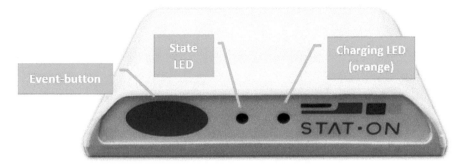

Figure 4.7 Sensor's interface.

- The drawings and specifications of the product and its components.
- Manufacturing requirements and procedures.
- Labeling and instructions for use.
- Design verification and validation, as well as chemical, biological, and functional testing, are performed according to applicable standards.
- Risk management report per EN ISO 14971.
- Clinical evaluation of the product following Annex X of directive 93/42/EEC and MEDDEV 2.7.1.
- The essential requirements checklist is in Annex I of directive 93/42/EEC.

Instructions for use
The user manual is a guide for the use of the sensor focused on the neurologist or the operator. It begins with a quick guide to installing the app and how to initialize the system, but it also explains the conditions of use, describes the parts of the system, the application, the outcomes of the sensor, and finally, it explains all the regulatory issues.

This document is extremely reviewed by the laboratories that certify the device. Thus, it is important that includes points such as the warnings, care and use instructions, indicating all the important information (who can use the device, its purpose, electrical isolation, contraindications, disposal instructions, and secondary or side effects, etc.).

The instructions for use document has to be readable, and a quick start is recommended.

In the STAT-ON™ case, the physical interface (Figure 4.7) is introduced, and the report is generated by the app when required, according to the registered and stored data (Figure 4.8 for details). In addition, a whole

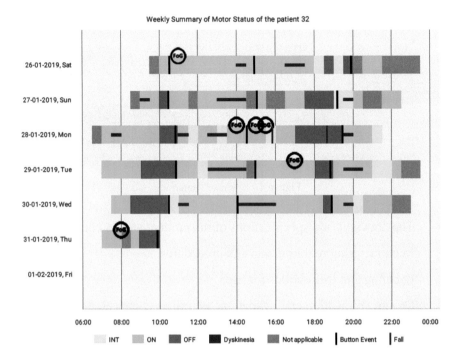

Figure 4.8 Weekly motor state report. The button pressed can indicate an intake of the medication.

section is included to explain the report to help professionals interpret it correctly.

A statement about data protection is also included, stating that in compliance with the general data protection regulation (GDPR), the company Sense4Care S.L. guarantees that collected data is uniquely stored within the device and that only the user is responsible for the use of these data.

In its present form, STAT-ON™ cannot share the collected data with a third party without the user's consent. Therefore, sense4Care S.L. will only access data under the express consent of the user and the owner of the STAT-ON™ device. Furthermore, shared data to Sense4Care S.L. will always be pseudo-anonymized and kept under the strictest security and confidentiality measures.

The technical documentation must include the complete list of the STAT-ON™ technical specifications and the ordered list of all the related standards and regulations affecting the device. This information was already included in Tables 3.7 and 3.8 of chapter 3.

4.3.2.3 STAT-ON™. Design and manufacturing information

This section of the technical documentation specifies the set of the necessary material to manufacture the STAT-ON™ device. Again, some electrical schematics and diagrams are included in this document as an annex. Moreover, specifications and regulatory tests must also be included.

The documentation includes the datasheets of the used devices. The datasheet of the used battery is particularly important in accordance with the IEC622133. The corresponding test report IEC62133-2 must also be included since this report guarantees the safety conditions of the battery and its use.

Another important document is the bill of the material, which is divided into two documents, the internal bill of material of the circuit, and the bill of material of the device.

Next, the production and final verification essays procedure must be attached, where the procedure is described, which the manufacturer must strictly follow. Finally, the batch file comprises the manufacturer procedures documents, including design and change controls, program elaboration, construction, replication, and installation procedures. Moreover, the fabrication orders, registers, and storage registers are also included.

The entitled "*Manufacturing and Verification & Final Tests*" document contains all the aforementioned information. It also contains all the manufacturing processes, including the following steps:

1. Fabrication order requested by the manufacturer.
2. Purchase of material.
3. Manufacturing order (subcontractor).
4. Phases of manufacturing (the SMD mounting and the assembly of the electronic elements in the enclosure of the equipment).
5. Review by the Responsible Technician (manufacturer).
6. Registration in the warehouse (subcontractor).
7. Shipment to the customer (subcontractor).
8. If it is in the subcontractor's warehouse for more than 1 month, it is sent to Sense4Care and stored in a locked cabinet at the Sense4Care offices.
9. Shipping to the customer (Sense4Care).

It is necessary to describe the inspection process. During the manufacturing process, the production personnel inspects the performed work (all the specific checks and inspections to be carried out are specified in the work instructions).

At the end of the operations of a job, the completion of the work performed and its verification will be recorded in the production records of the manufacturing order in process. In addition, in cases of incidents or defects, the losses caused are indicated in the same record.

Manufacturing records allow to the establishment of the quantities manufactured. The final inspections of the products are intended to determine if the product is suitable for marketing. To do this, a check is made with the model's specifications indicated in the manufacturing order.

In the event of noncompliant results of the inspections/checks, the entire manufacturing order will be rejected and will be treated according to the established nonconformity treatment procedure.

After reviewing and verifying the closure of the manufacturing order by the production manager and checking all the manufacturing and control records (batch file), in the event of favorable results, the technical manager releases the batch of products authorizing their placement, being available to commercial/sales. Otherwise, the technical manager retains the product, treated as a nonconforming product.

All records relating to manufacturing and release are filed together in the batch file to maintain the traceability of manufactured and distributed products.

The traceability and batch records are important and will be included in the quality system folder when the manufacturing process is executed every time. These files will control the number of units, purchase orders, and the number of nonconformities and will provide important information to the technical responsible for taking future decisions in the manufacturing process.

4.3.2.4 STAT-ON™. General safety and performance requirements

A systematic review of the fulfillment of the general safety and performance requirements/essential requirements set out in annex I of Directive 93/42/EEC must be provided in the technical documentation. The checklist indicates the applicability of each requirement, applied technical standards, and a pointer to the relevant sections of the technical documentation that support fulfillment.

A list of applicable/applied standards with the publication year is provided in the technical documentation. In addition, the list indicates whether or not the standards are harmonized with Directive 93/42/EEC.

Table 4.1 Risk management list for STAT-ON™.

Life-cycle phase	Risk management activity to be performed
Design	Identification of hazards and preliminary evaluation Risk control measures – selection and implementation Control measures – verification Preliminary Risks Management Report.
Transfer to production	Revision of the applicability and correct implementation of all the control measures that imply materials control, suppliers, subcontractors and/or process controls. Any unexpected risk or modified control measure will be documented and entered into the risk control records.
Routine production	Any abnormal tendency regarding the product safety characteristics will be analyzed to determine whether a corrective/preventive action is to be taken. The impact of the corrective/preventive measures taken to maintain or to increase product safety will be analyzed and entered into the risk management records.
Commercialization/ postproduction	Any customer complaint that originates changes/ corrections to the product will be analyzed to determine its impact on the existing risk evaluations. Feedback from the market will be monitored to determine whether it is convenient to implement corrective/preventive action.
End of useful life/ end of validity period	A device lifetime of 10 years (shelf-life) is scheduled for this product. However, product manufacturing samples will be kept in order to be able to confirm their functionality, even at the end of the specified device lifetime. Upon discontinuing product commercialization, this risks management plan will be closed, and all the documentation in the risks management file will be kept for a minimum of 10 years.

4.3.2.5 STAT-ON™. Risk management

Risk management is performed in each product life-cycle phase, following the requirements and activities set out in EN ISO 14971[3]. Generally, the life cycle defined for the product (for the case of STAT-ON™) is as indicated in Table 4.1.

The documentation that must be generated for the product and kept in the risk management file includes the definition of the risk management plan, the report of the initial risk management (with the risk assessment in the design phase and transfer to production), and the reports of the Product Review (including the continuous assessment of risks and including information feedback from the different stages of production and marketing routine/postproduction).

The methodology used for risk management activities is as indicated in the general procedure of "risk management," stated in EN ISO 14971, and includes the following items apply to all products:

- Establish the risk management policy and qualification of the team engaged in risk management.
- Definition of a scale of probability of occurrence of hazards and a range of levels of severity of the consequences if the hazard occurs.
- It establishes a general framework for risk acceptability criteria based on the combination of likelihood and severity levels.

For a given medical product, the following characteristics must be specified:

- Intended use and the features related to safety.
- Identification of hazards under normal and fault conditions based on the experience and the applicable regulations.
- Risk assessment (estimation of probability) associated with each hazard to determine initial acceptability.
- In the case of unacceptable risks, the analysis of possible causes or sources of hazards and the options available for controlling and/or mitigating the risk.
- Selection and implementation of the available options for controlling and/or mitigating the risk.
- Re-evaluation, after the implementation of risk control options and/or mitigation, to determine if there are residual risks.

The risk management report must include the following:

- The list of hazards is considered under both normal and fault conditions.
- Possible consequences for patients, users, and third parties and the possible causes.
- Estimating the initial risk (before control/mitigation), defining the implemented control measures/mitigation.
- Final risk estimates (after control/mitigation), including the determination of the acceptability of the final risk.

- Review the possible generation of new risks following implementing control measures/mitigation.

All the conclusions drawn regarding the effectiveness of the adopted measures and risk/benefit balance are set out in the final declaration of the report signed by the team responsible for risk management. As indicated in the EN ISO 14971, the following list of documentation is required:

- Risk management plan.
- Record of personnel qualification.
- Qualitative and quantitative characteristics.
- Dangers identification – risks preliminary estimate.
- Control methods/risk mitigation.
- Residual risk evaluation.
- Warnings to include in the label and instructions for use (IFU).
- Usability protocol.
- Validation of usability.
- Usability study report.

4.3.2.6 STAT-ON™. Product verification

In this part of the technical documentation, a set of documents must be included, according to directive EC 60601-1 [4], with the different published corrigendum and amendments, and the directives EN 300330V2.1.1 and EN 300328V2.1 [5, 6]: the obtained certificates, the performed laboratory tests, and others such as the transport test certificates. It must also include the clinical evaluation report, signed by experts, and the software validation, which is a set test performed on the app under the EN62304 regulation [7].

Software validation

The software validation part must include a series of documents that aims to test the firmware and the software associated with the medical device under EN62304. The tests carried out and validated by the manufacturer are also tested and validated by the notified body. This validation is carried out in order to provide documented evidence of the confirmation that the software app product gives correct and reliable functionality as per the legal and user's established requirements.

The identification of the person responsible for the performance of the validation process must be done and kept in the validation archive, along with the following documentation:

- Software validation plan records (validating personnel, source of the programs, flow diagrams, intended use, etc.)
- The related publications and reference standards and the applicable legislation and rules.

Since the STAT-ON™ product has two associated parts of the software (the Firmware and the user app), it is necessary to generate two separate software validation processes.

Clinical evaluation

The clinical evaluation report is a mandatory part of the product verification in the technical documentation of the medical device product. For the case of STAT-ON™, the followed methodology is according to directive 93/42/EEC (Annexes I.6a, II.3.1 and II.3.2. (c)) [8]. Furthermore, it must be considered that the evaluation of the clinical data is performed per Annex X, Section 1.1.1 of directive 93/42/EEC taking into account the guidelines set out in the guide "EU Medical Devices Documents" (MEDDEV 2.7.1 (rev. 4 of June 2016)).

The main objectives of the clinical evaluation are the following:

- Establish that the device requirements on safety (applied standards, etc.) are properly analyzed and that all the hazards, information on risk mitigation, and other clinically relevant information were identified and included in the information supplied by the manufacturer.
- Establish that the balance clinical benefit/risk ratio is positive when the system is used according to the established indications and purposes.
  - Any risks identified in the risk analysis are minimized and acceptable.
  - The intended purpose of the STAT-ON™ is supported by clinical evidence.
  - Warnings of residual risks included in the IFU are supported by sufficient clinical evidence.
  - Establish that the clinical benefits of the STAT-ON™ device are suitable within the widely accepted, given the current state-of-the-art, the intended use of the product, and the established clinical indications.
- Establish that undesirable side effects are acceptable.

- Establish provisions for proactive updating of the clinical evaluation to reflect changes in state-of-the-art through the application of postmarket surveillance (PMS) and postmarket clinical follow-up (PMCFU).

Considering the experience gained by the manufacturer along the redesign and development cycle of STAT-ON™, as well as relevant information from other similar products, the following actions have been taken in order to gather relevant information:

1. Searches of reference literature to establish the scientific basis widely established and technical needs that must be covered.
2. Search and review of the background and clinical experience, both published and unpublished, including comparison with similar or equivalent products already on the market.
3. Specific search of published relevant scientific literature and clinical research results focused on analyzing the possible benefits, safety, and injuries made by STAT-ON™.

In the case of STAT-ON™, strategies 1) and 2) have been used to establish the aspects and characteristics of the product for which the manufacturer believes that there is already sufficient scientific, technical and/or clinical data.

Strategy 3) has been used to study the clinical data related to the most relevant information to establish whether there is sufficient data concerning clinical safety and performance to ensure compliance with applicable regulatory requirements.

4.3.3 Part C: Updates/device modifications

This last part of the technical documentation is composed of documents that must be updated every year or 2 years depending on the requirement of the document in particular. This part is composed of the following:

- Change orders
- Technical documentation on postmarket surveillance
  - Postmarket surveillance plan (PMS plan)
  - PMS reports
  - Periodic safety update reports
  - Postmarket surveillance reports

The changes orders affect the STAT-ON™ device, for example, updates on the software app or modifications of the user manual, quality system, mechanical changes, or even electrical changes.

All these changes must be reported in the technical documentation and the quality system.

The technical documentation on postmarket surveillance must include information about sales, nonconformities, productions, planned actions, administrative information, and customers.

These PMS documents are filled out yearly or every 2 years with the information provided by the sales team and quality department. The objective is to establish a control on the number of sales for the management review and set the company plans for next year.

4.4 Quality Management System

This section describes the quality management system (QMS) of the company, which is based on the standard EN ISO 13485: 2016 [9]. The QMS is crucial for a company since it manages the quality of all processes and the products commercialized. Furthermore, the QMS documents the structure, procedures, responsibilities, and processes needed for effective quality management.

The QMS aims to manage all the processes that take part in the production and commercialization of a medical device, passing from the manufacturing of the device and the purchase of components to the sales control, customer satisfaction surveillance, human resources of the company, problems arisen, audits, company facilities, company structure, etc.

The major benefits of the QMS include the following:

- Enhancement of customer satisfaction by meeting their needs and requirements.

- Providing the right direction to achieve the company's objectives, goals, and mission.

- Maintaining and controlling documents and records.

- Helping in business expansion and growth.

- Identifying risks and generating opportunities to mitigate them.

- Improving the product and the process quality.

- Reducing cost and increasing productivity.

4.4 Quality Management System

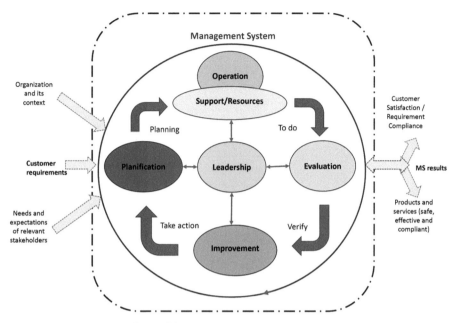

Figure 4.9 PDCA approach to the QMS.

- Engaging employees to achieve functional objectives and the organization's goals.
- Identifying and reducing process variations.
- Detecting and preventing defects or mistakes.
- Facilitating and identifying training needs of workers and staff.

Figure 4.9 shows a PDCA (plan-do-check-act) approach to the relationship between the main actors, processes, and expected outputs and results in the QMS of the company.

From a practical point of view, the most important document of the QMS is the quality manual, a real summary of the whole QMS. In this document, it is possible to find all the information about the company, the team, the facilities, the location, the aim and business of the company, the context in which it was created, the competitors, the scope and scope, and the relation with customers.

The QMS is established to implement the quality policy, to make the achievement of the quality objectives easy, and to ensure compliance with applicable regulatory requirements and with customer requirements. The QMS includes the policies, processes, and procedures to which reference is

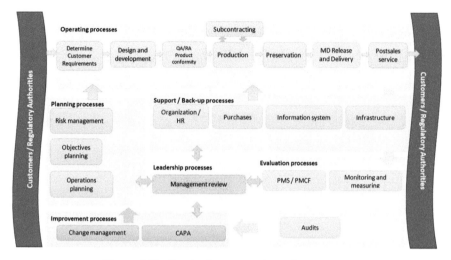

Figure 4.10 Involved processes in quality system.

made, the organizational structure of the company, and the precise responsibilities for the implementation of the activities to achieve the objectives set based on a focused approach in the process risk management document procedure.

The quality manual and the associated documentation establishes how to apply and maintain the QMS. Likewise, it identifies the criteria and methods required to guarantee the effective operation and control of the system. It identifies the measurements, monitoring, analysis, information, and actions necessary to achieve the planned results, conserve the system's effectiveness and improve it continuously.

Figure 4.10 shows the main processes necessary to implement the quality system for the manufacturing and supply of medical devices. The different interrelationships are also indicated.

As it is indicated, there exist an important set of processes related to the operative parts: determination of the customer requirements, the design, and development of the product, the production and manufacturing process, the conformity of the product with the regulation, the specific release and delivery process of the medical device, the postsales service, etc.

These processes must be planned (planification of the involved operations, how the risks are managed, and the objectives must be planned, etc.). There is a specific part for the evaluation of the processes with concrete measuring and monitoring processes, together with the required audits to be prepared periodically.

Additional processes for support are required: organization, purchases, information systems, and the necessary infrastructure. The processes for the improvement of the processes are also a very important part.

The quality system requires a very well-organized and structured set of documents. This is an important part of the work to be done and includes the following list:

Quality manual
It is a description of the quality system of the company. It contains references to the used documented procedures, the description of the QS processes, their interaction, and the critical reviews of the documentation structure.

Quality politics
It contains the mission statement of the organization as well as its intentions in relation to quality, risk management, and compliance with regulatory requirements.

Process map
It contains a diagram indicating the different relevant processes of the organization, the sequence, and their interaction.

Procedures
Describe the activities carried out in the framework of the QS to meet the requirements established in accordance with the reference regulations. Here, are also included the work instructions, the control guidelines / analytical methods, the manufacturing guidelines...

Quality records
They keep the results obtained along the quality process and show that the activities are carried out according to the applicable regulations and comply with applicable regulatory requirements.

Plans
They establish the necessary planning and programming to establish the time schedule to carry out the required actions to maintain compliance of the QS (audit plan, calibration plan, training plan, etc.).

Quality objectives
It contains the declaration of the concrete quantitative aspirations of the organization derived from the quality policy and related to quality processes, risk management, and compliance with regulatory requirements.

File of health product (medical device file)
It is necessary to maintain a file for each type of health product or family of medical devices, containing all the generated documents that are necessary to demonstrate compliance with the applicable regulatory requirements.

Process risk management file
This file contains all the reports and tests related to process risk management

Regulatory submissions
The file contains the documentation related to the regulatory presentations, such as the manufacturer license (to be presented to the AEMPS in Spain, the Register & Listing (to be submitted to the FDA in United States, etc.

Communications notified bodies
The file contains the documentation related to the notified bodies involved in the conformity assessment of the product (applications, certificates, audit reports, etc.).

Batch file/manufacturing records
These verified and approved record sets provide traceability of each product or many products. They also identify the quantities of manufactured and released products for commercialization.

External documentation
This is a set of reference documents. For example, the international technical standards and the legal requirements applicable to the products (regulation, standards/guides, contracts, agreements, etc.)

The different types of documents integrated into the structure of the QMS are described in the procedure structure and minimum contents of the quality system documents.

In the case of products manufactured by the company, the technical documentation of the product, according to the procedure preparation and control of technical product documentation, must be included (technical documentation already described in the above text).

In the case of imported/distributed products, the documentation will include the certificates, product documentation, and regulatory records according to the procedure file of imported/distributed health products.

It is important to note that:

- Maintaining registers/records of data is mandatory to provide evidence of the effective operation of the QS compliance with regulatory requirements and product and service conformity with the established customer and regulatory requirements. The registration forms must be included as annexes in the procedures that explain their use.

- The identification, storage, recovery, protection, retention time, and final disposition of the records are defined in the procedure of records control.

- The records can be both in paper and electronic format. Backup copies of the records are made in computer support. Records containing personal or confidential health data are duly protected per the applicable regulatory requirements.

- The records of the quality system shall be kept on file during the minimum period established in the QMS. This period will be longer than the product's useful life and at least 10 years (15 years in the case of implantable products) after the last product has been manufactured.

The documents listed above must be generated according to the corresponding procedures and instructions described in the QMS. In the case of a medical device, like the STAT-ON™ device, 60 items can be found, giving an idea of the work to be done and its complexity. Additionally, many of them include instructions and registers to be filled in regularly.

Since the complete list of the involved processes should be excessive, along with the rest of this section, a summary of the most representative processes will be presented according to the scheme of Figure 4.10.

- **Quality manual folder**

This is the most important part of the QMS, where the quality manual is included, and the structure of the company is described. It also includes: the list of documents, the involved processes and their relationship, the responsible for each of them, the quality policy, and the designation by the board of management of the different charges and roles, etc. Figure 4.11 shows the typical structure of a company manufacturing medical devices.

- **Description and procedure management**

In this procedure, the main objective is to establish the necessary information for the planning, management, and monitoring of the processes used by the company to carry out its activities in the frame of the QMS.

110   *The EU Medical Device Regulatory Process*

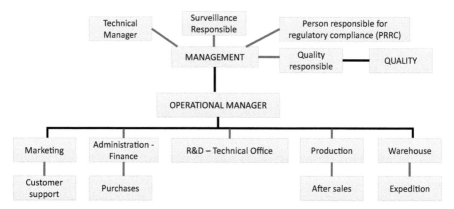

Figure 4.11 Company's structure as described in the quality manual.

- **Manage outsourced processes**

In the case of Sense4Care (the company manufacturing STAT-ON™) a great part of the production is subcontracted. The management outsourced processes are very important since it describes the relevant processes of the company that are outsourced or subcontracted as established in the process map (operating processes part). The processes establish their control and maintenance to confirm their continuous suitability for the company. It also includes the treatment of interactions with crucial suppliers/suppliers (suppliers that provide materials or services that may have a critical and direct impact on the conformity of the final marketed product).

Those involved in the control of the process and the owner of the process will maintain the control of the process monitoring, alerting about the quality by opening a corrective/preventive action, in case of observing an anomalous trend.

- **Process risk management**

The management of risks is a crucial aspect of the QS, and it is implemented as part of the QMS. The risks associated with the procedures can affect the activities of the company, coming from any internal or external source. They can also come from outsourced processes. Additionally, all the applied changes in current practices may also have an impact, generating a risk.

For this reason, managing risks is not an activity limited to certain people or areas of the company, and it is generally applicable in all areas.

This procedure, as part of the planning processes, establishes a methodology by which all types of risks must be proactively detected and managed to try to minimize their impact and the actual or potential damage they may cause according to EN ISO 31000.

- **Software quality validation**

Embedded software or necessarily related applications are important parts of the actual medical devices. Therefore, the procedure's main objective for the software validation is to establish the methodology used for the validation before initial use and the subsequent periodic revalidation of the software through the preparation and execution of master validation plans.

This procedure applies to all computer applications (software) used in the quality management system and in the product being the software app and the internal firmware of the microcontroller in the case of STAT-ON™.

- **PS file – preparation and control of technical documentation**

The medical device file, also called technical file under the MDD or technical documentation under the MDR, must follow some structure, rules, and control of the documents. The structure is described in Section 3 of this document. This procedure determines the methodology and systematics for preparing the medical devices' technical documentation (the technical documentation – TD). In the case of STAT-ON™, this file is important and according to the specification given in the technical documentation above.

- **Documentary file-Fab**

This procedure is an important piece of the Operating processes since it defines the system that allows tracing the materials used in the manufacturing process of the marketed products, both internally and by our subcontractors, as well as the destination of all the medical devices manufactured by the company and the period of conservation of the records associated with these said activities.

In order to maintain the company's activity within the quality standards and according to the regulations, it is fundamental to organize the review and planification objectives. The following processes must be implemented with this objective:

- **QMS planning** (yearly definition of improvements)

- **QMS review** (definition of the necessary information for the periodic review of the quality system)

- **Risk management product** (development of the risk management policy adopted by management to guarantee the supply of medical devices that are safe, effective, and fit for their purpose, keeping the potential risks associated with their use acceptable in relation to the benefit for the patient and compatible with a high level of safety and protection of health, considering the general current knowledge)

- **Design and change control** (all medical devices could be affected by some changes, in the app, in the hardware, or even in the QMS. It is important to notify and follow a set of rules such as the procedures to identify the problem, the modification performed, and the gravity of the modification, and notify the NB if this is relevant for the MD)
- **Internal audits** (determination of the conformity or nonconformity of the quality management system with the planned and documented provisions, with the requirements preestablished by the company and the applicable regulations. Analysis of the effectiveness of the quality system. Satisfaction of the regulatory requirements of surveillance and prevention of nonconformities)

STAT-ON™ is a medical device Class IIa with the EC mark. The QMS of the manufacturing company (Sense4Care) must implement and follow a set of processes to comply with the applicable regulation and legislation. The most important processes are:

- **Determination and monitoring of requirements** (established strategy for regulatory compliance and regulatory requirements applicable to products). It includes:
    - the steps to be taken to determine and periodically review the applicable regulatory and normative requirements.
    - description of the steps to follow to check the regulatory requirements applicable to the product in the countries and jurisdictions where it will be placed on the market and how to proceed with the conformity assessment.
    - description of the steps to follow for the registration of new products (in Europe, in Spain, etc.)
- **Qualification and classification of products** (establish a method to determine if a product falls within the scope of the regulation applicable to medical devices. If this is the case, establish a method of determining the risk classification of the product according to the applicable regulation. It also defines a method to determine the category and subcategory of the product, and a method to determine the naming code applicable to the product, etc.).
- **Product registrations – EUDAMED** (EUDAMED is the main database of the European Union concerning medical devices. The procedure defines the nomenclature of the product and the process of product registrations in EUDAMED).

4.4 Quality Management System

- **Clinical evaluation and postmarketing clinical follow-up (PMCF)** (establishment of the procedure for systematic searching, appraisal, and analysis of relevant clinical data, including the review of the postmarket surveillance (PMS) clinical experience and the postmarket clinical follow-up (PMCF) of medical devices, by planning, conducting and documenting a clinical evaluation according to the requirements established in Article 61 and Annex XIV of regulation (EU) 2017/745).

- **Conformity assessment** (it documents how to proceed with the Conformity Evaluation of the devices with intervention of a notified body, specifically in the stages of precommercialization and postmarket).

- **Software (SW) manufacturing and installation** (definition of the actions taken to carry out the construction, copies, installation and maintenance of the final version of a medical device (medical software). Establishes the mechanisms used in the production of a computer program that runs on a Smartphone or tablets).

- **Distribution of medical devices** (defines the protocol for the medical device distribution: contracts, agreements, requirements for each country, etc.)

- **Identification – UDI** (establishes the process of treatment for the UDI for Europe. The UDI is the unique device identifier for the medical devices registered in Europe).

As STAT-ON™ is a medical device mainly sold in Europe, and for this reason, it is necessary to define a set of processes to define and ensure the quality of the postmarketing and the relationship with the customers:

- **After-sales service and technical assistance** (establishes an after-sales service that ensures the customers the correct installation and maintenance of the products sold).

- **Customer feedback (postmarket surveillance)** (definition of a proactive and systematic postmarket surveillance system that includes: analysis of production information of the product, product tracking, early detection of observed adverse events, detection of opportunities for product and safety improvement, etc.)

- **Compliance with regulatory requirements** (establishment of compliance with the product's applicable market requirements, including those established by customers and those established by regulations).

- **Customer satisfaction** (establishment of the process for obtaining and using information related to the client's perception, regarding compliance with the established requirements. Each manufacturer must decide which is the correct customer contact channel).
- **Claims treatment** (definition of the mechanism to treat any claim done by the customer. The main goal is to ensure that all claims received are dealt with in a diligent and expeditious manner and ensure that claims are analyzed to determine if an incident must be reported to the competent authorities. The following types of claims are contemplated: product operation and features, product safety, reliability and duration of the product, identification and/or appearance of the product, packaging, labeling defects, etc.)
- **Regulatory audits and inspections** (within the yearly performed audit, it is necessary to establish the QMS for receiving regulatory audits/inspections from notified bodies and/or sanitary/competent authorities).
- **Treatment of noncompliant product** (establishment of the methodology to try to solve appearing problems and nonconformities with the product. The company must ensure the identification and control of any material, component, process, or product/service that does not comply with the applicable requirements and prevent its unintentional use or delivery. Finally, it is determined the registers that must be kept to indicate the nature of the nonconformities found and the actions are taken to correct them).

In order to ensure success in the implementation of the company's QMS, it is very important to have the active involvement of the staff and diverse personnel. Therefore, among others, the following processes must be implemented:

- **Human resources and competence** (definition of the competences of each one of the managers and staff of the company. Establishment of the responsibilities and authorities associated with the different jobs. It is also established how the necessary competency for people performing work that may affect product quality is defined and reviewed).
- **Staff training and education** (definition of how staff training needs are identified, how training actions are planned, and the information to be recorded for the training actions that are carried out).

4.5 Conclusions

The regulatory process of a medical device is a complete and complex process that involves several players. The immensity of the file structure and the rigorous surveillance performed by authorities make this process very expensive, time-consuming, and exhausting. On the other hand, this process's rigorousness is the reason why medical devices are safe for patients, reliable for medical healthcare professionals, ensuring that no cheap, easy, or dangerous device could affect the integrity of any patient.

As shown in the present chapter, there are three main important parts in a regulatory process: the manufacturer's license, the technical documentation, and the quality management system. Different authorities audit all these parts and the final outcome is a single certificate called EC certificate. The complete process can take a long time (around two years, according to the personal authors' experience), highly depending on the saturation of the regulatory system and the available notified bodies.

Currently, Europe is living in a period of several changes with the introduction of the new MDR, which is valid from May 2021, the Brexit process that obligates the manufacturers to achieve a new certificate called UKCA, and the global crisis scenario given by several factors are affecting the plans of the companies. Furthermore, this whole situation is saturating the notified bodies' activity, enlarging the time for auditing, and being dangerous for the roadmaps of companies to meet the time requirements of customers and project deadlines.

Apart from the required time, the related costs might vary depending on the laboratory tests a medical device requires. After this, the quality manager and technical manager are responsible for keeping the system notifying the notified bodies and authorities when necessary. The QMS is audited yearly and requires a systematic expense that is difficult to keep.

STAT-ON™ was certified in June 2019, and so far, successive yearly surveillance audits have been done. The QMS has already been adapted to the MDR; now, the technical documentation must be adapted to the requirements of the European Commission in the coming future. It must also be considered that the EC certificate achieved by Sense4Care expires five years after the date it was achieved, as well as the manufacturer's license. The main advantage is that since the technical documentation is audited every year, only a few changes will be required for the renewal.

The chapter presented details on the process to be followed in Europe. Similar steps must be done in territories such as Japan or United States with the PMDA and FDA, respectively. The process practically begins from the beginning, although the QMS and technical documentation are similar. The TGA in Australia is another certificate, but as European Union and Australia have some agreements, the process is easier. Other territories are of interest, but the company must establish a trade-off between the expenses of the team in the regulatory process and the benefits achieved in each territory, making it a challenge for the company to reach new territories.

Sense4Care roadmap is to completely adapt the sensor to the new MDR, achieving as well the UKCA, the TGA, and jumping to the United States and Japanese territories by achieving FDA and PMDA certificates, respectively.

References

[1] Regulation (EU) 2017/745 of the European Parliament and of the Council of 5.4.2017
https://eurlex.europa.eu/legalcontent/EN/TXT/PDF/?uri=CELEX:32017R0745&from=ES

[2] IEC 62133-2:2017 Secondary cells and batteries containing alkaline or other non-acid electrolytes – Safety requirements for portable sealed secondary cells, and for batteries made from them, for use in portable applications

[3] UNE-EN ISO 14971:2020 Application of risk management to medical devices.

[4] IEC 60601-1:2005 Medical electrical equipment – Part 1: General requirements for basic safety and essential performance

[5] EN 300330V2.1.1 Radio equipment directive

[6] EN 300328V2.1.1 Wideband transmission system directive

[7] UNE-EN 62304:2007/A1:2016 Medical device software – Software life-cycle processes

[8] EC Council Directive 93/42/EEC of 14 June 1993 concerning Medical Devices

[9] UNE-EN ISO 13485:2016 Medical devices – Quality management systems – Requirements for regulatory purposes

5

STAT-ON™: The User Interface and Generated Report

Daniel Rodríguez-Martín[1], Carlos Pérez-López[1,2], Martí Pie[1], and Albert Pagès[1]

[1]Sense4Care S.L. – Cornellà de Llobregat, Spain
[2]CSAPG – Consorci Sanitari de l'Alt Penedès i Garraf, Research Department, Spain

Email: cperezl@csapg.cat; (daniel.rodriguez) (marti.pie) (albert.pages)@sense4care.com

Abstract

STAT-ON™ is a medical device capable of detecting and measuring the most relevant motor symptoms of Parkinson's disease. This device must be easy to use, both by the patient who must wear it during his normal activity, during the period prescribed by their neurologists, and also by themselves who, based on their interpretation of the measurements provided, will be able to know much better the state and evolution of the disease, in the treated patient.

In order to facilitate the use, an accurate conception and design of the appropriate user interface were done. This interface is based on a physical part (which allows direct interaction with the sensor), and a software part (in the form of an app that must be installed on a smartphone) that allows a series of interactions to get, at the end of the monitoring process, a report on the patient's condition. This chapter describes the details of the STAT-ON™ user interface.

5.1 Introduction

The complete STAT-ON™ system comprises a monitoring device, its base charger, a belt, and a mobile application. The system provides numerical and graphical information on the motor symptoms associated with Parkinson's disease. Furthermore, data related to the general motor activity of the patient are calculated.

The device continuously collects the inertial signals associated to the patient's movement, processes them in real-time using artificial intelligence algorithms, and stores the results in its internal memory. The sensor must be only managed in clinical environments, and only health staff can operate the app and the device. Therefore, the patient should wear the sensor in their daily activities to provide relevant information to health professionals.

The available smartphone application (the app) connects to the STAT-ON™ device via bluetooth (BLE). The app is used both for configuring the system and downloading the data previously generated by the sensor. In addition, the mobile application can send the data enclosed in a report by e-mail or digital support to any user, caregiver, therapist, or neurologist.

In previous chapters, it is possible to get all the details concerning the internal electronic and processing structure of the device, how it operates, and how the required regulatory process has been followed to obtain the CE marking as a medical device Class IIa. Along this process, it has been already introduced the necessity for a specific software (the app) that will be the necessary interface with the user. Through this software, it is possible to configure the STAT-ON™ correctly and, later, when required by the user to get the results stored in the device's internal memory, which will be processed under the format of a useful and understandable report.

The following sections present the requirements of the implemented software, the details about the functionality and related interface for an easy user experience, details about the generated report, and some hints on how to correctly interpret the contained information.

5.2 Requirements, Interface Description, and Different Modes of the Device

STAT-ON™ is a device designed to be very useful and easy to use. The patients must wear the device, and some interaction could be required from them by the neurologist (be aware of the correct position, to check the battery life of the sensor, to indicate the moment of the medication intake by pushing the device's button, etc.). From their side, the neurologists and healthcare professionals are the main users of STAT-ON™ and should enter the configuration according to the patient to be monitored, and at the end of the testing period, they should obtain the detected and registered information in relation

5.2 Requirements, Interface Description, and Different Modes of the Device

Table 5.1 Description of the main user requirements.

User requirement	Description
1	The interface must be easy to use and understand.
2	The app must run on Android and IOS smartphones and tablets.
3	The user must to obtain the app by downloading it from the official channels (Google and Apple stores).
4	The interface must be available in different languages.
5	The connection with the STAT-ON™ will be via bluetooth low energy (BLE). This connection must provide access to the results of the monitorization and to the status data.
6	The app must supply error messages to the user whenever needed and will always inform the user of the current status of the connected sensor.
7	The app will be responsible for the generation of a useful report containing all the organized information detected during the testing period. The user (health professional) is only responsible for managing and sending this information.

to the motor symptoms of the patient. This information should be useful and correctly structured for a correct understanding of the information captured by the sensor. Thus, the device's user interface is a very important part and must be carefully designed.

STAT-ON™ is equipped with a physical interface responding to the already discussed electronics presented in Chapter 3 (a press-button and a set of LEDs are part of this human–machine interface – HMI, permitting obtaining internal information about the operation modes and the state of the battery, and/or to enter a signal by pressing the button that indicates an event, like the moment of the medicine intake). Accompanying this physical part, a software interface has been included for the rest of the functionality and a correct interaction with the device. A very important component is the app to be installed in a smartphone, providing a complete human–computer interface (HCI). This software interface should permit the configuration of the device, according to the patients' characteristics, and give information about the state of STAT-ON™ (battery level, connection, etc.). When required by the neurologist, the HCI must be able to present, in an organized and useful way, all the information about the detected and measured PD motor symptoms.

Table 5.1 shows a minimum user requirement list that the implemented user interface must accomplish.

120   STAT-ON™

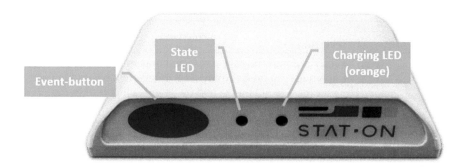

Figure 5.1   Physical interface.

5.2.1 The physical interface (HMI)

The sensor device has a button, and two led indicators next to the STAT-ON™ logo. The device is also equipped with a small vibrator motor and a buzzer (see Figure 5.1 for details).

- **The button**: The functionality of the available button is just for three specific cases:
  - Turn on the sensor when it is in shutdown mode.
  - Stop an alarm after it triggers.
  - Mark user events when specified by the professional (medication intake, sleep, meals, etc.).
- **The state LED indicator**: The color pattern of the State LED specifies the current status of the device. Through a sequence of different blinking colors (black, white, blue, magenta, green, and red), the LED shows the different states of STAT-ON™ (see reference [1] for complete details). The main possible states of the device are the following:

Connected and low battery indication
The sensor will indicate that it has an active Bluetooth connection or a low battery level by blinking the led in blue or magenta color, respectively. These indications will be combined with the sensor's current main state.

Shutdown
The sensor will come in this state initially. While in this state, the sensor will do nothing until its button is pressed. To power it up, place the sensor on its charging

pad and make sure the charging process starts (the orange led must switch on), then wait until the battery is fully charged (the orange led switches off). Then, press the sensor's button, and it should enter *CONFIGURATION_PENDING* state.

In addition, the sensor will automatically enter this state if the battery level is too low, to power it up, **the button should be pressed after charging the sensor**.

Configuration pending

When the device is in this state, its status LED will blink in white color. The device will not record data nor execute algorithms while in this state. In order to leave this state and start monitoring, the user should configure the following parameters: patient ID, age, leg length, and Hoehn and Yahr value. These should be configured through the STAT-ON™ app via Bluetooth.

Once the sensor is configured, it will alternate *SLEEP* and *MONITORING* states, which are the normal operation states.

Monitoring, sleep, and standby

When the sensor is correctly configured and has detected some movement, it enters the *MONITORING* state. The patient's movement is monitored in this state, and the algorithms are executed. In addition, the status LED will blink in green color. This normal operation state implies that the sensor is running correctly. However, if no movement is detected for some minutes or the sensor is charging, the device may enter *SLEEP* state in order to save power. The device will resume monitoring after detecting any movement.

Given that the power save mode is enabled and disabled automatically; **the user does not need to power the device on or off**.

The *STANDBY* state is an optional state that can be enabled once the sensor is correctly configured. It can be enabled using the <Standby> button in the configuration area in the app. This option forces the sensor to pause monitoring without losing its configuration. Once the sensor's button is pressed, the sensor will resume monitoring.

Full memory

If the internal memory of the device fills up, its status led will blink in red color. Since there is no space in memory, the sensor will not record any new data. It is therefore recommended to synchronize the device data using the STAT-ON™ app. After the data is sent, the device memory will be automatically cleared, and the sensor can monitor again. Formatting (clearing the sensor's memory) can also be done, but in this case, the stored data not yet synchronized will be completely lost.

Synchronization

The synchronization process involves transferring the stored data from the sensor to the smartphone. This can only be done by using the STAT-ON™ app. While this process is ongoing, the status led will quickly blink in blue, and the app will show a progress bar. After receiving all the data from the sensor, the app will automatically generate the corresponding files and reports (see the following sections for details).

Format

The format process completely clears the device memory. Formatting the sensor is only recommended if the device will not be used for a long time. Synchronizing the data contained in the sensor is recommended before starting the format process; otherwise, all the stored data not yet transferred to the smartphone will be lost. After formatting the device, its previous configuration will also be lost; thus, the sensor must be configured again to re-enable. The format sequence can be started by using the app and pressing the <DELETE> button.

Error

If the sensor detects an internal system malfunction, it will enter *ERROR* state. The status led will stay in red color. Most processes and operations, like monitoring or executing algorithms, are interrupted if an error happens. When the sensor connects with the app while in an error state, it will transfer the error code to the app, and the app will offer to perform a sensor reset.

- **The charging LED**: This LED indicates when the device is in the correct charging process, if the process had a problem, and when the charging process has finished. See reference [1] for more details. Figure 5.2 shows the correct use of the charging platform.

5.2.2 The sensor modes

The STAT-ON™ has different operational modes: shutdown, configuration, off, on, and sleep. After unpacking the system, the sensor device will stand in shutdown mode. Before using the sensor for first time, it is necessary to fully charge the battery and press the sensor's button to switch it on.

Once the button is pressed, the system will enter the configuration mode, from which the user can configure the sensor with the app. Then, the system will work autonomously. That means the user will not have to switch it on or off.

5.2 Requirements, Interface Description, and Different Modes of the Device

Figure 5.2 Device correctly placed on the charging platform (LED always orange while charging and always off when the charge is completed).

The system will enter sleep mode if there is no movement for some minutes. Then, it will automatically exit this mode and start monitoring after movement is detected. This work mode enables saving energy, thus extending the autonomy of the sensor.

If the user expects not to use the sensor for long, keeping it in shutdown mode is recommended. Shutdown mode is activated after formatting the device using the STAT-ON™ app. It is recommended to synchronize all the data before formatting in order not to lose all the data stored in the sensor permanently. Charging the device's battery before switching it off is also important.

In the regular use regime, the system works autonomously, that is, the patient does not need to interact with the device. The health professional will provide the sensor to the user correctly configured, and the user will wear the sensor for registering the symptoms of PD during the days of the study proposed by the health professional.

The healthcare staff can ask the caregiver to press the button at a certain time, such as lunch, dinner, medicine intake, etc.

The patient should use the system (worn on the waist) for a minimum of 5 days and 24 h within these 5 days to generate enough inertial data to personalize the algorithms.

It is recommended to use the sensor for 7 days. From this moment, a report can be generated at any time (see next sections). The doctor will download to

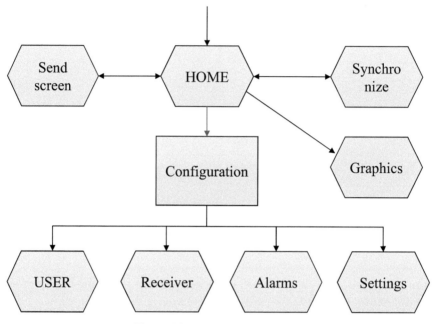

Figure 5.3 Software architecture.

his/her mobile phone the information generated by the sensor at the doctor's office with the STAT-ON™ application, which will automatically generate a report of the motor state and symptoms during the days of study.

After this step, the sensor will enter the initial state, configuring the required parameters to start a new study period with a new patient.

5.2.3 The software interface (HCI)

The software interface is provided through a specifically designed app to be downloaded and installed in a smartphone. This app must follow the concrete user requirements described in Table 5.1 and should respond to the structure shown in Figure 5.3.

The app was designed to have a home screen from which it is possible to do the following actions:

- Synchronize with the sensors.
- Generate graphics for visualization.
- Send data.
- Configure the device.

The configuration menu is used to set up all required parameters for the proper function of the app and consists of four screens:

- User screen to introduce the user parameters configuration.
- Receiver, to set up the information of the receiver of the generated data.
- Alarms to set up the alarms of the device.
- Settings, for the configuration of the sensor's parameters.

In order to create a multi-platform and multi-device app, the graphic elements must accomplish different requisites. Image quality and resolutions must be well designed and defined to achieve the best quality and good app performance without memory leaks. Appropriate tools have been used for these purposes. The app design was directly implemented by XML in both IDEs (Android Studio layouts and XCode interface builder), accomplishing different screen dimensions and devices. Some elements were also adapted to new design rules by Google in Android (Material Design) and iOS limitations or suggestions by Apple.

5.3 The Application (App) and Its Management

The STAT-ON™ app can be installed on any smartphone or tablet running Android 5 or higher, and the device must support bluetooth low energy (BLE) and have a 1GB RAM minimum. It also works in iOS for Apple devices. It is required to use iOS 10.2 or higher. The app can be downloaded at *Google Play* (Android) or the *App Store* (iOS), it must be searched for "STAT-ON," and make sure its developer is Sense4Care.

The STAT-ON™ device is suitable for evaluating the motor state of a patient with PD. The value of the *"patient ID"* item, which can be set through the app's *Configuration Area*, is used to associate all the data related to each user. There is no limit to the number of patients registered by the smartphone at the same time, it depends on the memory of the smartphone; however, it is recommended to use no more than six patients.

The **patient ID number must be changed each time a sensor is given to a different patient**.

In order to simplify the situation where a single user (usually a healthcare professional) handles various sensors and multiple patients, **the results and reports are obtained solely from the data transferred during the current synchronization event**. Therefore, no historical record is kept inside the app's database (the data monitored is used for generating the reports and then discarded and not used anymore). **However, in the Android version, the app does store all the generated reports** (.pdf and .csv) inside the STAT-ON-specific directory in the smartphone memory. Given that the generated reports are tagged using the patient ID number, it is still important to keep a value for each user and configure the sensor accordingly.

5.3.1 HOME: Main screen

After opening, the app shows the main screen, which enables access to all the areas and features of the STAT-ON™ app (see Figure 5.4). It also indicates whether there is an active connection with a sensor and shows its battery level. While on the main screen, the app connects automatically to the paired STAT-ON™ sensor. When a device is connected, it is announced by Bluetooth and battery indicators (see the top of Figure 5.4). "*Connected*" appears under the Bluetooth logo, and the battery level is also shown.

From this home screen, the professional (user) is able to perform and organize the basic functionalities:

- *Bluetooth:* Establish the correct Bluetooth pairing with a given device.

  The Bluetooth area manages the paired devices and chooses the sensor to connect to. Bluetooth has to be enabled for the app to connect with the sensor.

  Below the Bluetooth switch, the currently connected sensor is displayed, if any. Only one STAT-ON™ sensor should be connected at the same time. In order to search for the connectable sensors, the scan button should be pressed. Then all the available sensors will appear inside the area below. The device's Bluetooth name starts with *Stat-On* and then contains the two last digits of its serial number (e.g., "*StatOn00*").

  It is necessary to press on the STAT-ON™ device we wish to pair to, and a PIN/Passkey request may pop up. Each device has a six-digit numerical PIN/Passkey, which is provided with the sensor packaging.

- *Configuration:* Properly configure the device according to the patient data.

  The values inside the configuration area are stored inside the sensor. Thus, an active Bluetooth connection is required for its use. Once the user configures all the parameters, the user has to push the <SAVE> button.

5.3 The Application (App) and Its Management 127

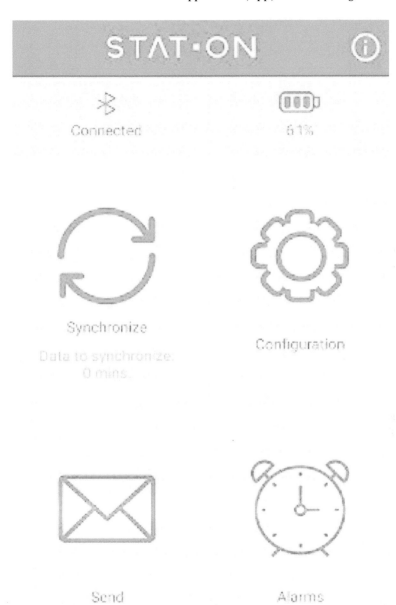

Figure 5.4 Main screen aspect.

Figure 5.5 Configuration menu.

If the sensor has any results from previously stored monitorizations (i.e., not synchronized), changing some parameters from this area will not be possible. Synchronizing or deleting the pending results is required before changing the parameters. In reference to Figure 5.5, it can be mentioned:

- ○ The patient ID value identifies the patient that wears the sensor. Patient ID is key for keeping the record of each patient correctly related and must be modified each time the sensor changes from patient to patient.

- o The age, the Hoehn and Yahr value, and the leg length of the patient must be introduced.
- o The SAVE button must be pressed after the introduction of the above parameters. This way, the configuration parameters are sent to the device and start the monitoring process. A green blinking light appears in the sensor Status LED.
- o The STANDBY button is available only when the device is correctly configured and permits to pause the monitoring process of the sensor.
- o The DELETE button permits clearing all the data currently stored in the sensor, whether it has been sent to the app by synchronizing or not.
- o The RESET button resets the sensor. This option is used in case the sensor blocks or has an error. It does not lose any data.
- o TURN OFF the sensor switches OFF totally. The timestamp is lost, but the internal data is not lost. It might be used when STAT-ON™ is not used for a long time.

- *Synchronization:* This option enables downloading the information that has been computed in real-time in STAT-ON™. With this option, once downloaded, the app generates the report and the CSV file (see Figure 5.6).

This option can only be used when connected to a sensor. The menu generates a basic five-page or extended report with all the daily information.

There is also an option to adjust the date and time of the desired monitoring period. This is used, for example, to delete data at the beginning or at the end of the monitoring period, which is not useful (i.e., the sensor has been sent through the post office, and there are 2 days of useless data). This way, the report generated only contains important information.

The synchronization button enables the download of all the sensor's data.

When synchronizing, all the results from the sensor are transferred to the smartphone using Bluetooth. This screen also shows the last time a synchronization had been performed.

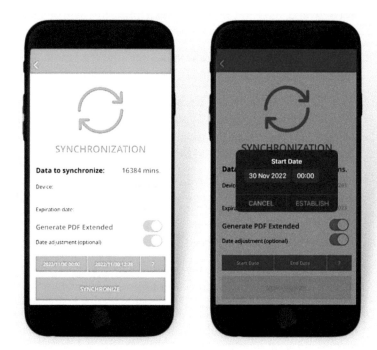

Figure 5.6 Synchronization menu,

- *Send:* Establish the sending functionality of the app.

  When the user synchronizes the device and downloads all the data, the app directly allows sending the information to a digital platform. However, if the user desires to send this information to another user later, then the "Send" option must be used. It acts like the common "Share buttons," so the user can choose any other communication app (like e-mail) for transferring the documents. A copy of the generated documents is also stored inside the mobile device's storage, under the <STAT-ON> folder. At the bottom of the screen, the <SEND> button will open a standard "share" dialog. Any mailing or file share method can be used.

- *Alarms:* It permits the optional configuration of alarms to be received on the device. The alarms will be stored and will trigger on the STAT-ON™, not the smartphone. When an alarm triggers, the sensor will vibrate until the sensor's button is pressed. If the "Sound" switch is enabled using the app, the sensor will beep, too. For example, alarms can be used to remember the patient's medication intake.

5.3.2 The reports

5.3.2.1 Introduction

When required by the professional, it is possible the generate a complete report about the patient's status. The application offers the possibility of generating two kinds of reports:

- A *Basic report* of the patient's condition and some graphics that condense the behavior of the symptoms and some gait parameters. The purpose of this first kind of report is of ordinary use for conventional clinical practice.

- An *Extended report* includes a large part of the parameters extracted from the algorithms and the information contained in the first type of report. The main purpose of this second mode of reports is its use in the field of research or for a more accurate analysis of the condition.

In order to ease the use and the correct understanding of these reports by professionals, graphic representations are extensively used. Therefore, the data are presented in four sections.

- Summary page
- Distribution and severity summaries
- Weekly summaries
- Daily information

The information provided in the report is the following:

- ON state
- OFF state
- INT state
- Dyskinesia
- Number of freezing of Gait (FoG) episodes
- Duration of FoG episodes
- Stride fluidity (bradykinesia index, BI)
- SMA (quantity of movement)
- Falls

- Events (indicated by pressing the button)
- Number of steps
- Step length
- Cadence

5.3.2.2 Summary of the STAT-ON™ measurements

Various symptoms associated with the patient's motor states can be differentiated in PD. One of the most common clinical practices is visually analyzing how patients walk in order to evaluate bradykinesia. In the activity of walking, several symptoms converge with different origins within the neurophysiology of PD. In gait, two movements of different natures are coordinated. On the one hand, automatic movements are classically associated with symptomatology related to hypokinesia, and on the other hand, voluntary movements that are associated with bradykinesia. It should not be forgotten that the pathophysiology of bradykinesia is the cardinal symptom per excellence of PD. Furthermore, this symptom has a greater degree of correlation with the level of dopamine deficiency and, therefore, with the fluctuations between motor states in PD. Peak-dose dyskinesia is a side effect of the medication that clearly indicates the patient's motor status associated with the ON state.

FoG is another symptom that is of special interest because it is one of the most disabling symptoms of PD. In addition, FoG has different characteristics from other Parkinsonian symptoms; for example, it has not been possible to clearly correlate the frequency of FoG episodes with other motor symptoms of PD, such as stiffness and bradykinesia. Although, in many cases, it is not a particularly useful symptom to assess the patient's motor status, it is useful to evaluate the evolution of this symptom and the mobility difficulties of the patient.

The detection method of ON/OFF states in patients with PD depends on the characterization of the motor symptoms that the patient presents in each state. In this sense, two specific detectors are used, which analyze the presence of dyskinesia and the bradykinetic gait. The outputs of the detectors are merged into a global classifier that estimates the motor state.

The bradykinesia detector is based on the analysis of patients' gait and has been validated in several studies that can be found in [2]–[5]. Since this detector is self-adaptive, it must have a minimum data period of three days. From this analysis, an important index is shown in the reports called stride fluidity or bradykinesia index, which is correlated with subscales of the UPDRS concerning bradykinesia and gait [6, 7].

The detector of choreic dyskinesia is mainly based on detecting the frequencies of dyskinesia maintained during prolonged periods of time. The outputs of these algorithms are combined through a decision tree, which performs the detection of the motor states. The detail of these algorithms can be found in [8].

The presented architecture has implications for interpreting the data presented in the graph. The most relevant is that the sensor emits an OFF verdict when the patient walks. In other words, in those patients with very deep OFF states in which they cannot move, STAT-ON™ will not be able to issue a verdict. On the other hand, ON states are associated with prolonged physical dyskinesias in time, in addition to the bradykinesia level.

As presented in [3], since the bradykinesia algorithm is self-adaptive, another implication is that **the system will only show this information if a minimum of 3 days of data has been captured.**

The FoG detector is based on the analysis of windows of 1.6 s; therefore, this is the minimum temporal resolution. This means that although freezing episodes lasting less than 1.6 s are detected, all of them will be reported as 1.6 s long. Another example can be that two episodes of 1.8 and 3.1 s will be notified as episodes of 3.2 s. This means that when STAT-ON™ reports a FoG episode of 1.6 s, it will last from 0 to 1.6 seconds, whereas when a 3.2 s episode is reported, it will result in a duration between 1.6 and 3.2 s. For more details on this detector, go to [9].

It must be noted that the total number of reported falls might be confused since the system also analyses the movements when the patient removes the sensor belt or puts it on. These moments involve movements that could be similar to a fall and the system could generate a false positive. The detection of activities, and more specifically, the length and speed of the step, are algorithms specifically developed and adjusted with data from patients of PD. Details of this group can be found in [10]. Below, a detailed description of each of the graphs and data generated by the STAT-ON™ system is presented.

In the following sections, the different parts of the basic and extended report are presented and discussed. Every section title announces the content, also mentioning if corresponds to the basic or extended reports.

5.3.2.3 The summary page (basic and extended report)

The report's summary page presents a series of numerical data as a summary of the physical activity of the patient and the prevalence of symptoms that the patient has presented during the monitored period (see an example in Figure 5.7).

User ID:	4
Age:	72
Hoehn & Yahr:	2.0
Study start:	21/02/2022
Study end:	01/03/2022
Days monitored:	9
Time Monitored:	97 hours

Nº FoG Episodes:	28
Average FoG Episodes/day:	3.1±2.3
Average minutes walking/day:	87.4±38
Average number of steps/day:	9783.6±4389.1
Motor inactivity (% time monitored):	20.5 hours (21.1 %)
Total Time in OFF (% time monitored):	35.5 hours (36.6 %)
Total Time in Intermediate (% time monitored):	16.5 hours (17 %)
Total Time in ON (% time monitored):	24.5 hours (25.3 %)
Total Time with dyskinesias (% time monitored):	27.5 hours (28.4 %)
Bradykinesia index (Stride fluidity) >8.5 optimal; <6.5 suboptimal	6.6±0.4

Figure 5.7 Summary page example.

In the first table, the specific data from the patient and the monitored period is shown:

- User ID: Numeric identifier of the patient, introduced through the app by the professional.
- Age: Age of the patient.
- Hoehn and Yarh: PD stage evaluation.
- Study start date: Day and Hour of the start of the monitored phase.
- Study ending date: Day and Hour of the end of the monitored phase.
- Total days monitored: Total number of days the patient has been monitored.

In the second table, a summary of the symptoms and physical activity during the monitored period is shown:

- *Total FoG episodes*: Total number of FoG episodes that have been measured during the monitored period.

- *Average FoG episodes per day*: It is a comparable relative measure between patients or separate monitoring periods. Standard deviation is also provided, which gives evidences as to whether the patient has FoG episodes consistently every day or whether there are days that show more than others.

- *Average minutes walking per day*: It is a good indicator of the physical activity presented by the patient.

- *Average number of steps per day*: In patients without gait disorders, it provides very similar information to walking minutes, but in the case of presenting gait disorders, this parameter is significant to assess the disease.

- *Time in OFF (% regarding total time monitored)*: Percentage of time monitored in which the patient presents OFF state.

- *Time in intermediate (% regarding total time monitored)*: Percentage of time monitored in which the patient presents an INTERMEDIATE state.

- *Time in ON (% regarding total time monitored)*: Percentage of time monitored in which the patient presents ON state.

- *Time with dyskinesia (% regarding total time monitored)*: Percentage of time monitored in which the patient has evidenced dyskinesia episodes.

- *Bradykinesia index (stride fluidity)*: this index represents the patient's state after the monitored period. It is considered that an index below 6.5 is considered a patient in a suboptimal state. Conversely, a patient who is over 8.5 is considered a patient in an optimal state.

5.3.2.4 Symptoms distribution graph (basic and extended report)

One of the most relevant graphs presented in the report is the weekly representation of the patient's motor symptoms. An example is shown in Figure 5.8.

The daily time is included on the horizontal axis, while the monitored days are indicated on the vertical axis. The colors in the graph represent the different states of the patient according to the following code:

- Green: The patient is in ON state.

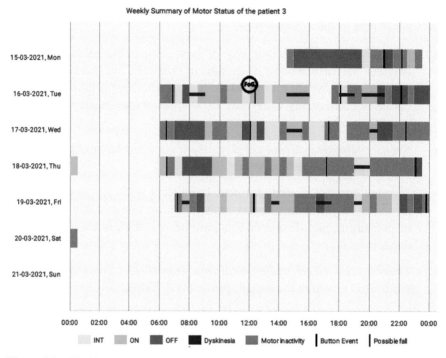

Figure 5.8 Weekly motor state. The presence of FoG is indicated. In this case, the button was pressed at medication intake.

- Red: The patient is in OFF state.
- Yellow: The patient is in an intermediate state.
- Magenta: It has been detected as choreic dyskinesias.
- Gray: No state has been detected (no dyskinesias, no walking detection).
- FoG circle: Detection of FoG episode.
- Blue vertical line: indicates a possible fall.
- Black vertical line: indicates an event. The patient pressed the button in that moment. It might indicate that the patient has taken the medication. The clinician might suggest the patient do so to see the correlation between the symptoms and the medication effect. It can be used for other purposes such as eating, feeling bad, falling, or sleeping.

5.3 The Application (App) and Its Management

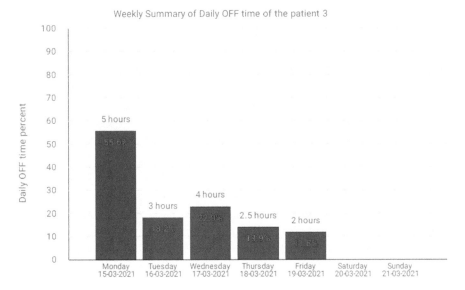

Figure 5.9 Time in OFF state.

5.3.2.5 Graph of the weekly time in OFF state (basic and extended report)

This graph (see an example in Figure 5.9) shows the daily-accumulated time in OFF every day and the percentage of OFF time regarding the total time monitored every day.

On the horizontal axis, the days monitored (maximum 1 week) are shown, and on the vertical axis is the percentage of monitored time that the patient has been in OFF state. Although the bars are based on the percentage of monitored time detected as OFF state, information is added about the number of hours the patient has been in this state. Whenever this graph is analyzed, three factors must be taken into account: the total monitoring time, the sum of hours in OFF state, and the total time with any motor state verdict. Analyzing this graph jointly with the weekly motor state is highly recommended.

5.3.2.6 Graph of the weekly FoG episodes (basic and extended report)

On the horizontal axis (see Figure 5.10), the days monitored (maximum 1 week) and the number of episodes detected per day can be observed on the vertical axis.

138   STAT-ON™

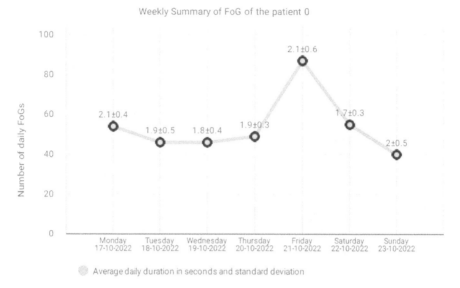

Figure 5.10   Weekly FoG episode detection.

This graph shows the number of episodes detected per day, the average length (as explained above in a resolution of 1.6 s), and the maximum duration of an episode of FoG per day.

5.3.2.7 Graph of the weekly stride fluidity (basic and extended report)

The graph presents the weekly evolution of the median stride fluidity that the monitored patient presents. The arrows represent the best and worst fluctuations of the patient throughout the day. The personalized thresholds for the ON and OFF are also indicated with a red and green line. The average of these two numbers is the bradykinesia index parameter found on the summary page. The red and green zone indicates the mentioned optimal and suboptimal zone. When the patient is over 8.5, it is considered an optimal motor state. However, when the patient is below 6.5, the patient is considered to be in a suboptimal zone. This graph is very useful for seeing the patient's daily fluctuations and severity. It permits comparing with other patients and the same patient after a medication adjustment.

The stride fluidity is a measure of acceleration obtained as an intermediate result of the bradykinesia detector (ranging from 2 to 25), which is related to the fluidity of the patient's movement when walking. This way, this evaluates the evolution of the patient's difficulty when walking as an average per day (the greater the value, the greater the fluidity).

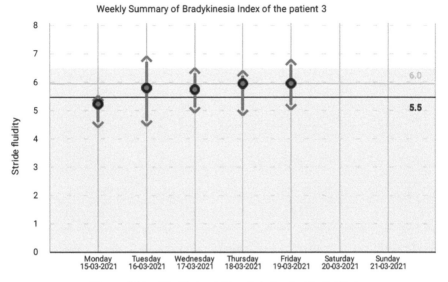

Figure 5.11 Weekly bradykinesia index (stride fluidity).

This value is correlated with the so-called factor 1 of UPDRS-III (see [2]). Figure 5.11 shows the days monitored (maximum 1 week) on the horizontal axis and on the vertical axis the measure of fluidity.

5.3.2.8 Clinical interpretation guideline (basic and extended report)

At the end of each report, basic or extended, a quick guideline is presented that indicates some of the ranges for considering the warnings of the summary page (see Figure 5.12). Given that some graphs might provide a lot of information, this guideline is useful as support for the clinician to interpret the patient's state with more quality.

5.3.2.9 Graph of physical activity (extended report)

In the extended report, the sensor also gives data about the physical activity that the patient has performed during the entire monitoring period. The measured variables shown are:

- Step length
- Stride speed
- Cadence

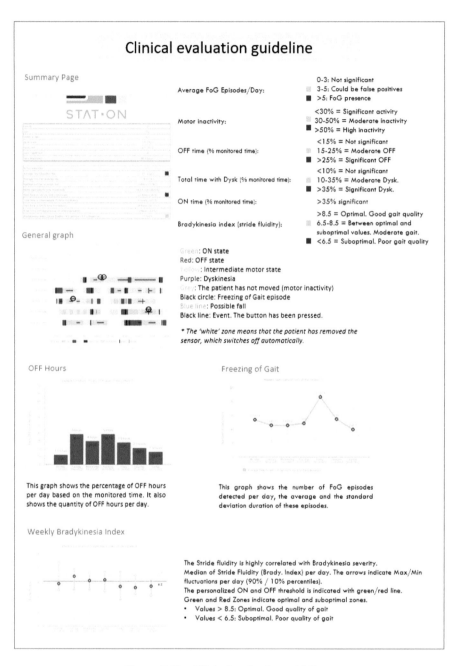

Figure 5.12 Clinical evaluation guideline.

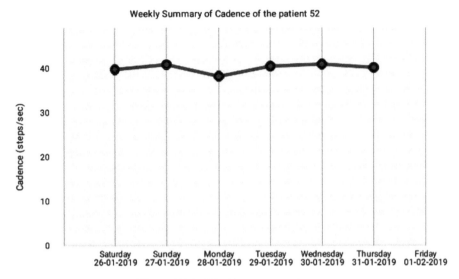

Figure 5.13 Weekly average cadence.

- Quantity of movement
- Number of steps

In each of the related graphs, it is shown, on the horizontal axis, all the monitored days, and on the vertical axis, the average per day of the units corresponding to each one of the measurements. For example, Figures 5.13 and 5.14 show a couple of examples.

5.3.2.10 Graph of daily motor symptoms (extended report)

STAT-ON™ generates a graph of motor symptoms per monitored day where it can be seen, in addition to the motor status, the dyskinesia occurrence, and the number of FoG episodes the patient has suffered, informing the hours of appearance. The resolution in all the daily charts corresponds to half an hour. Figure 5.15 shows an example.

On the horizontal axis, it is shown the hours of the day, and on the vertical axis, a series of labels that describe the corresponding row:

- Time Monitored: Time at which the sensor is running.
- ON/OFF/INT state: representation of the motor state detected in the patient. Red corresponds to OFF state, Green to ON state, and yellow to the intermediate state.

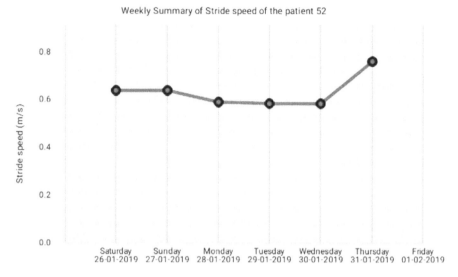

Figure 5.14 Weekly average of stride speed.

- Dyskinesia: periods in which choreic dyskinesias have been detected in the patient.

- FoG episodes: The number of FoG episodes is represented in this row. If an FoG episode is detected, a box with the number of episodes is drawn.

5.3.2.11 Graph of daily stride fluidity (extended report)

The system generates a graph of the stride fluidity when the patient is walking, where the daily evolution of the stride fluidity of the patient's gait can be assessed.

In addition, in the background of the graph, the detected motor state is also drawn (red for OFF, green for ON, and yellow for INT). Finally, note that the thresholds calculated (by a self-adaptive algorithm), upper (green) or lower (red), are also drawn. These thresholds indicate when the patient is in its OFF zone or ON zone. The threshold changes between patients, given that the sensor learns how the patient walks. The thresholds are set based on machine learning methods and establish the patient's ON and OFF zones. This graphic is very interesting in understanding how the patient fluctuates and how much fluctuates.

On the horizontal axis, it can be observed the hours of the day and on the vertical axis, the units correspond to the stride fluidity (m/s^2).

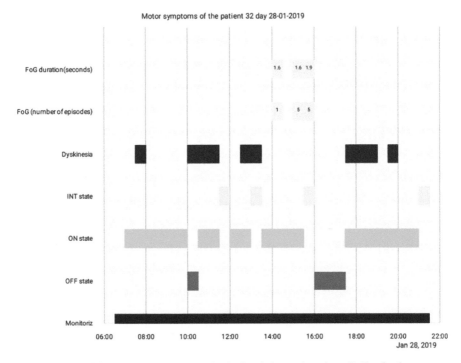

Figure 5.15 Daily motor states, including information about FoG episodes.

5.3.2.12 Graph of daily physical activity (extended report)

This group of indicators provides detailed information about the physical activity the patient has performed throughout the day and during the days they have been monitored. These variables are:

- Step length
- Cadence
- Energy expenditure
- Number of steps

In each of the graphs, the hours of the day are shown on the horizontal axis, and the units corresponding to each measurement are shown on the vertical axis. Figure 5.17 shows the daily energy expenditure as an example.

As it was also introduced above, the system can produce a reduced version of this complete report. The reduced report is just a selection of the information and graphs contained in the extended report.

Information with special interest in order to help the clinical professionals to have a more complete and objective view of the state of the PD

144 STAT-ON™

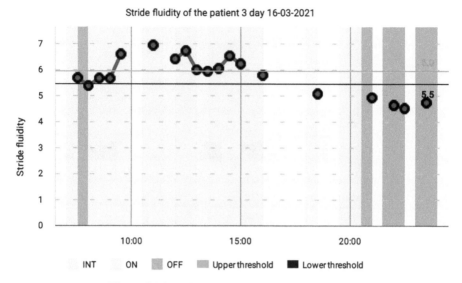

Figure 5.16 Daily stride fluidity and motor states.

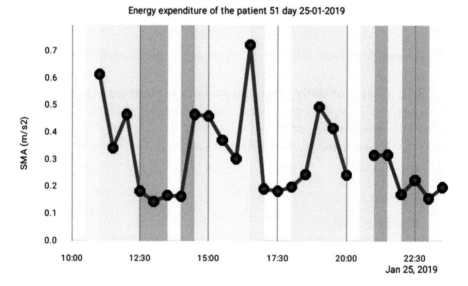

Figure 5.17 Daily energy expenditure and the motor states in the background.

patient, was selected. In concrete: the Summary page (Section 5.2.2.3), the weekly motor state (Section 5.2.2.4), the weekly time in OFF state (Section 5.2.2.5), the weekly FoG episodes (Section 5.2.2.6), and the bradykinesia index weekly graphic (Section 5.2.2.7).

5.4 Report Hints and Interpretation

As a part of the user (mainly the health professional) interface of the STAT-ON™ device, the contents of the extended report have been presented in previous sections. This report and the graphical representation of the captured and measured data was determined according to the list of the users' requirement, shown in Table 5.1, and after many discussions and professional opinions from several cooperating neurologists.

The measured data of the different relevant motor PD symptoms and the calculations and algorithms applied to them, in order to be stored in the device, and represented under the form of a report when required from the app, is explained and presented in [9].

The objective of this section is to try to clarify some of the contents of the report to make their correct interpretation easier for the proper use of this information (and the contained data) for the correct management of the PD patient.

The following text presents some hints and comments on the interpretation of some parts of the report. This is not a complete list, but some specific aspects have been identified:

5.4.1 Some interpretations on the weekly summary of motor state graph

We refer to the graph presented in Section 5.2.2.4. It is necessary to remember that green color corresponds to the ON state, red to OFF, yellow corresponds to the intermediate state and gray means that the period is not applicable since the patient is not walking. The magenta line represents the presence of dyskinesia, the black line corresponds to pressing the button, and the blue line indicates the occurrence of a fall.

Figure 5.18 corresponds to a couple of days of the complete monitored period of a patient presenting some interesting particularities:

- It can be seen how the patient has presented dyskinesias at some moments of the monitoring, but that he **presents an OFF-motor state during a great part of it**.

- On the other hand, it presents some isolated FoG episodes, which means that, during the half-hour of resolution that the graph has, at least 1 FoG episode has been detected.

- In this representation, several conditions must be taken into account. One of the most important is that the **diagnosis of OFF can only be**

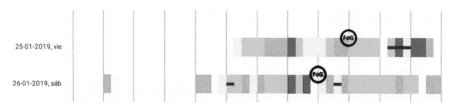

Figure 5.18 Part of the weekly summary graph of a patient.

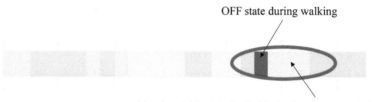

Figure 5.19 Example of active periods followed by inactive ones.

issued when the patient is walking due to the functioning characteristics of the sensor based on accelerometers. Therefore, in moments when a verdict cannot be issued, the monitoring zones without diagnosis appear (in gray).

- Some patients do not walk or walk less during episodes of deep OFF. This means that these **episodes in gray may be long periods of inactivity, or the patient is so bad that he cannot even walk**.

Another interesting aspect to be discussed can be seen in Figure 5.19, where the patient walks a lot throughout the day, and although he presents intermediate states, he continues walking. However, when it goes into an OFF state, he stops walking for a long time.

When the system issues the diagnosis again, it is already in ON state again. We can appreciate how the OFF state is only counted for half an hour, although it may take longer (remember that the sensor can only detect and measure when the patient is moving). In these cases, **it is important to consult with the patient if, during the OFF periods, he tries to walk or not**.

A comment and explanation must be made on the fall detection capabilities of STAT-ON™. It is convenient to consider Fall detection as an indicative mark since it can generate some false positives in specific situations such as when the patient lies down in bed or on the sofa very abruptly or when the patient leaves or puts on the seatbelt due to the manipulation movements.

5.4 Report Hints and Interpretation   147

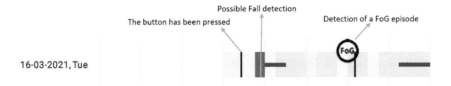

Figure 5.20 Some details related to fall detection, the existence of FoG, and pressing the button situation.

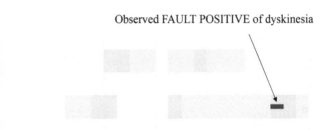

Figure 5.21 A false positive was observed in a healthy person wearing STAT-ON™.

This is a possible consequence of not receiving any feedback from the user (the patient wearing the system).

Figure 5.20 shows a day distribution of a given patient where FoG appeared at a given moment, and a fall has been indicated during an ON state, and at a given moment, the patient pressed the button.

Concerning the presence of FoG, this kind of graph shows that in the half-hour slot around the indicated circle, at least one FoG episode has occurred. For more details and information, it is necessary to consult the specific information detailed in section 5.2.2.6.

The button press mark (black vertical line) appears each time the patient presses the button. This functionality can be used as desired and according to the indications done by the neurologist. For example, in the concrete study shown in Figure 5.20, the user was asked to press it when he took medication.

During the validation of the system, monitoring tests were carried out with healthy patients, obtaining profiles of activity. It should be noted that on certain occasions (such as shown in Figure 5.21), there may be false detections depending on the activity that is taking place (in this case, it is a bus trip).

In healthy individuals, false positives of some symptoms (especially dyskinesia and FoG) have been observed when performing activities such as scrubbing the floor, cleaning the oven, **or taking some public transport such as a train or bus**. *It is relevant to advise patients to try to carry out* their activities of daily living normally, but to remove the sensor if they are going to carry out physical activities (e.g., going to the gym) or activities with highly repetitive movements such as scrubbing the floor. It is suggested to clinicians that they ask the patient to fill in a simple diary explaining if they take some transport or do sport.

5.4.2 Some details on the weekly FoG episodes

This comment is referred to the detection of FoG episodes and their weekly representation (see Figure 5.10). On the horizontal axis, the monitored days (maximum one week) can be observed, and on the vertical axis, the number of episodes detected per day. This chart shows the number of episodes detected per day, the average duration (as explained in Section 5.2.2.6 at 1.6-s resolution), and the standard deviation of the duration per day.

FoG detection is based on the movement patients make when they are blocked, which can be similar to some movements carried out in physical activities such as the gym or classes of various dance disciplines. **Therefore, it is necessary to ask the user if they have carried out these activities during the monitoring process and to recommend that they remove the sensor when they are going to carry them out**. Furthermore, in the same way as the previous section, it is recommended to fill in a simple diary explaining if they perform some activity such as sport or taking some transport.

5.4.3 Some recommendations for a correct use of STAT-ON™

Due to the internal design of the device, based on accelerometry and the treating algorithms for the detection of PD-related motor problems, it is very convenient to be aware of some basic recommendations.

These recommendations are formulated in base on the acquired knowledge during the design process and the user experience in real cases:

- The STAT-ON™ system is a human movement analyzer; therefore, it is completely discouraged to carry it for long periods in public and private transport as it can cause false positives in some detectors.

- Patients should be advised to try to carry out their activities of daily living usually, but to remove the sensor if they are going to perform physical activities (e.g., going to the gym) or activities with highly repetitive movements such as scrubbing the floor.

- The patient may be required to press the button to indicate any circumstances the neurologist indicates (e.g., taking the medication, thinking he is entering an OFF period, etc.). He should, therefore, avoid pressing it in any other circumstance.

5.5 Conclusion

The chapter has presented the complete user interface for the correct use and reported measurement understanding. For more detail, it is recommended to access the STAT-ON™ user manual [1] or the product's website [12].

References

[1] STAT-ON™ user manual. Available from Sense4Care website.
[2] C. Pérez-López et al., "Assessing Motor Fluctuations in Parkinson's Disease Patients Based on a Single Inertial Sensor," *Sensors*, vol. 16, no. 12, p. 2132, Dec. 2016.
[3] A. Rodríguez-Molinero et al., "Analysis of correlation between an accelerometer-Based algorithm for Detecting Parkinsonian gait and UPDRS subscales," *Front. Neurol.*, vol. 8, no. SEP, pp. 3–8, 2017.
[4] A. Rodríguez-Molinero et al., "A Kinematic Sensor and Algorithm to Detect Motor Fluctuations in Parkinson Disease: Validation Study Under Real Conditions of Use," *JMIR Rehabil. Assist. Technol.*, vol. 5, no. 1, p. e8, Apr. 2018.
[5] À. Bayés et al., "A 'HOLTER' for Parkinson's disease: Validation of the ability to detect on-off states using the REMPARK system," *Gait Posture*, vol. 59, no. September 2017, pp. 1–6, 2018.
[6] Rodríguez-Molinero et al. "Analysis of correlation between an accelerometer-Based algorithm for Detecting Parkinsonian gait and UPDRS subscales". *Frontiers in Neurology*, 8(SEP), 3–8 (2017). https://doi.org/10.3389/fneur.2017.00431
[7] Samà, A et al. "Estimating bradykinesia severity in Parkinson's disease by analysing gait through a waist-worn sensor". *Computers in Biology and Medicine*, 84. (2017) https://doi.org/10.1016/j.compbiomed.2017.03.020
[8] C. Pérez-López et al., "Dopaminergic-induced dyskinesia assessment based on a single belt-worn accelerometer," *Artif. Intell. Med.*, vol. 67, pp. 47–56, Feb. 2016.

[9] D. Rodríguez-Martín et al., "Home detection of freezing of gait using support vector machines through a single waist-worn triaxial accelerometer," *PLoS One*, vol. 12, no. 2, 2017.

[10] T. Sayeed, A. Samà, A. Català, A. Rodríguez-Molinero, and J. Cabestany, "Adapted step length estimators for patients with Parkinson's disease using a lateral belt worn accelerometer.," *Technol. Health Care*, vol. 23, no. 2, pp. 179–94, 2015.

[11] J. Cabestany, A. Bayés (editors), 'Parkinson's Disease Management through ICT: The REMPARK Approach' River Pub. 2017.

[12] statonholter@sense4care.com

6

STAT-ON™: The Holter for Parkinson's Disease Motor Symptoms. Real Use Cases in Clinical Praxis

Núria Caballol*[1,2], Angels Bayés[2], Anna Planas-Ballvé[1],
Tània Delgado[3], Asunción Avila[1], Alexandra Pérez-Soriano[2,12],
López-Ariztegui Núria[4], Sònia Escalante[5], Diego Santos-Garcia[6,13],
Jaime Herreros[7], Iria Cabo[8], Jorge Hernandez-Vara[9],
José Maria Barrios[10], Lucía Triguero[10], Alvaro García-Bustillo[11],
and Esther Cubo[11]

[1]Departament de Neurologia, Complex Hospitalari Moisès Broggi, Sant Joan Despí, Spain
[2]Unitat de Parkinson i Transtorns de Moviment, Centro Médico TEKNON, Barcelona, Spain
[3]Hospital Parc Taulí, Sabadell, Spain
[4]Unidad de Trastornos del Movimiento, Hospital Universitario de Toledo, Spain
[5]Hospital Verge de la Cinta, Tortosa, Spain
[6]CHUAC – Complejo Hospitalario Universitario de A Coruña, Spain
[7]Hospital Universitario Infanta Leonor, Madrid, Spain
[8]Complexo Hospitalario Universitario de Pontevedra, Spain
[9]Departamento de Neurología, Grupo de Investigación Enfermedades Neurodegenerativas, Campus Universitario Vall d'Hebrón, Barcelona, Spain
[10]Unidad de Trastornos del Movimiento, Hospital Universitario Virgen de las Nieves, Granada, Spain
[11]HUBU – Hospital Universitario de Burgos, Spain
[12]Fundació de Recerca Clinic Barcelona, Hospital Clinic de Barcelona, Spain
[13]Departamento de Neurología, Hospital San Rafael, A Coruña, Spain
(*) Chapter coordinator

Abstract

As part of the introduction and diffusion strategy of the STAT-ON™ device among neurologists and professionals, treating patients affected by Parkinson's disease, a number of experiences have been promoted in several hospitals and movement disorders units.

The present chapter explains 13 real use cases experimented in Spanish hospitals, mainly in the period 2020–2021, using the device to help the professionals with current treatment activity. In the totality of the presented cases, advantages have been obtained with the use of STAT-ON™. Details are reported in the following sections.

6.1 Introduction

This chapter presents a series of real clinical use cases developed over the last 2 years, including the use of the STAT-ON™, as part of the care and treatment process provided to various patients with Parkinson's disease.

As can be seen in the affiliation of the different authors, the collaborating hospitals and involved movement disorder units and services are distributed throughout the Spanish geography. These selected experiences are part of the introduction and dissemination process of the new STAT-ON™ technology in the regular neurologists' medical activity to treat already diagnosed patients with Parkinson's disease.

The previous chapters presented the complete process through which important scientific and technological results, materialized in the REMPARK project, became a class IIa medical device ready to be introduced to the market. This process means that a series of challenges and requirements of the process itself has been covered (scientific-technological quality, safety of use, adaptation to existing regulations, etc.) leaving, nevertheless, a very important aspect for this type of product, which consists of its acceptance for being used in medical practice by neurologists and professionals. It is very important, therefore, that STAT-ON™ can prove its usefulness, as a complementary technology, when used in the patient care process to provide relevant and decisive information for improving therapies.

The chapter has been coordinated by Dr. Núria Caballol and the authors, responsible for each use-case, are referred at the beginning of each section.

An attempt has been made that the different cases have a similar structure, which focuses on the presentation of the personal histories and those referring to the Parkinson's disease of each one of the patients. A presentation of their condition is made according to the physical examination carried out, which in many cases (when the doctor considers it necessary) includes the result of various applied scales and tests. This is followed by the objective pursued, for each case, with the use of the STAT-ON™.

Results obtained are presented, showing relevant parts of the reports generated by the device, which allows addressing a series of discussions and conclusions related to each specific case, and which try to highlight the benefits obtained through the use of the technology. In many cases, this opens the door to being able to carry out an improved screening process, an improvement in the treatment, a better therapeutic adjustment, or the advanced identification of a series of characteristics related to the disease.

As a relevant conclusion, it is worth saying that the use of the STAT-ON™ device contributes to a better knowledge and understanding of the disease by the patients, helping them to be more aware of their condition (representing, also, an increase in the patients' empowerment level).

It is necessary to mention that, as the use cases are real, some of the presented figures, containing parts of the STAT-ON™ generated reports, are in Spanish since this is the language used by the authors in their contributing hospitals.

6.2 Early Detection of Motor Fluctuations

Responsible professional: Dr. Angels Bayés.
Unitat de Parkinson. Centro Médico TEKNON. Barcelona.

Personal history: This 65-year-old man worked as a commercial director in a data protection company. He has the habit of doing sports frequently and intensively. In 2013 he interrupted his sports activity due to a knee injury. Past clinical history includes olfactory dysfunction since many years, viral pericarditis in 1986, and surgically treated meniscopathy.

Parkinson's disease history: In 2013, some changes in the sleep pattern began in the form of fragmented sleep. Since May 2014, he has presented difficulties in writing with alterations in neatness and size. In addition, he referred to clumsiness for fine motor skills, such as picking for coins in a pocket and a slight resting tremor in the left hand.

Dat-Scan was performed in November 2014, which showed bilateral putaminal hypoperfusion, being worse on the right side. He was diagnosed at this time with Parkinson's disease. In 2015, he lost 5 kg of weight in 6 months. The disease has evolved slowly throughout these years, with progressive clumsiness for both automatic and voluntary movements and muscle rigidity, especially affecting fine motility. In addition, he has presented progressive difficulty in verbal communication, drooling, and low mood, with a tendency to self-social isolation. In March 2016, he started a low dose of levodopa (100 mg/day) with improvement in sleep and tremor. The dose of levodopa was progressively increased to a current dose of 450 mg/day, associated with opicapone. The patient reports a good motor response to levodopa and no fluctuations in relation to medication, although he reported tiredness/fatigue, especially in the afternoon.

At that moment, he was taking the following medication: rasagiline (1 mg 1-0-0), pramipexole (2,1 mg 1-0-0), levodopa/carbidopa (100/25 mg 1-1-1-1 (intakes at 7-12-17-22-24 h), and opicapone (50 mg 0-0-0-1).

Physical examination: Left-handed. Well-oriented patient, with good cognitive status and no motor or sensory deficits. Mild stiffness of the neck and upper extremities, predominantly on the left. Mild fine motor disability, predominantly on the left side. UPDRS (test performed on March 17, 2022): Mental activity: 1; activities of daily living: 8; motor exploration: 14; Hoehn and Yarh: II; Schwab and England: 70%.

STAT-ON™ objective of use: Given the suspicion of motor fluctuations, a study with STAT-ON™ was proposed. He was asked to wear the sensor for 12 h a day for 7 days and told to press the button on the device after each levodopa intake.

Diagnosis and decision-making: According to the STAT-ON™ record (see Figure 6.1), the patient presented motor fluctuations. A delayed ON was detected in the first levodopa dose in the morning and in sporadic doses, especially in the afternoon. Some dyskinesia was also detected, and it must be indicated that the patient was not aware of it. Some freezing of gait episodes were detected, specifically during OFF or wearing-off periods.

Given the patient's complaint of being suboptimal, with evidence of delayed ON/wearing-off, as reported by the STAT-ON™, an increase in 50 mg of levodopa was indicated at 5 pm. After this adjustment, the patient-reported experiencing a generalized better clinical state.

Discussion: It is well-known that there exists a lack of awareness of most people with Parkinson's disease. This also includes the difficulty in detecting early ON–OFF fluctuations. This data is crucial to adjust the treatment as soon as possible and improve daily quality of life.

Conclusions and take-home messages: The use of STAT-ON™ has been very useful in verifying that this patient has motor fluctuations and making him aware of them, helping to adjust the treatment more precisely.

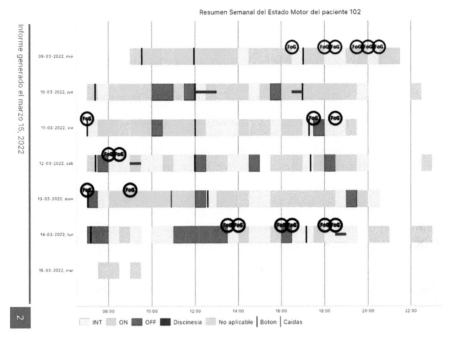

Figure 6.1 Weekly summary reported by STAT-ON™ (it shows delayed-ON in the morning, between 7 and 8 am). Additional OFF periods appeared around 12 am, and later, between 5 and 6 pm were also detected.

6.3 Improving Awareness of the First Motor Fluctuations

Responsible professional: Dr. Anna Planas-Ballvé
Complex Hospitalari Moisès Broggi (movement disorders unit) in Sant Joan Despí. Barcelona.

Personal History: He is a right-handed, Caucasian, 59-year-old man, ex-smoker, working as a telecommunications engineer with a history of mild obstructive sleep apnea.

Parkinson's disease history: His symptoms started at 53 years of age with rest tremor, stiffness, and fine motor clumsiness of the right upper limb, impairing his daily activities and tasks at work. The patient also reported olfactory loss since he was approximately 45 years old. Laboratory workup including complete blood count, renal, liver, and thyroid function was normal. Magnetic resonance imaging of the brain was unremarkable. A diagnosis of clinically established Parkinson's disease was established, according to the International Parkinson and Movement Disorder Society (MDS) task force criteria.

At 54 years of age, treatment with an extended-release form of ropinirole (8 mg/day) optimally controlled the Parkinsonian symptoms. One year after, symptoms were bilateral but markedly asymmetric, and carbidopa/levodopa (75/300 mg) was started with positive control of symptoms.

However, a few months later, safinamide (100 mg/day) was started because the patient explained mild general disability, and dragging of the right leg when walking was observed. In addition, the patient did not notice wearing-off or dyskinesia. At that time, the Hoehn–Yahr (HY) stage was 2, the UPDRS-III score was 10, and the levodopa equivalent daily dose (LEDD) was 590 mg.

STAT-ON objective of use: The use of the device was proposed since the neurologist suspected initial motor fluctuations and with the objective to assess the progression of the first motor fluctuations.

Diagnosis and decision-making: According to the information provided by the sensor, morning akinesia was detected almost every day, and the percentage of daily time in OFF was 4.2%. However, the patient denied having morning akinesia or wearing-off.

Surprisingly, the sensor detected freezing of gait (FoG) in ON, intermediate and OFF states. However, the patient denied suffering from true FoG episodes. Due to the fact that the patient dragged his right leg while walking, it was considered that the device was detecting this motor phenomena of "unilateral dragging gait" (see Figure 6.2)

Due to the results of the first STAT-ON™ report, the carbidopa/levodopa dose was increased (112.5/450 mg), and opicapone (50 mg) was started. In addition, due to bothersome leg edema, dopamine agonists were discontinued gradually.

At 57, the patient began noticing morning akinesia that lasted approximately 15 minutes and wore off motor symptoms in the afternoon. Another STAT-ON™ monitoring period was indicated, with the same HY stage and UPDRS-III as in the first record but higher LEDD (675 mg). The Holter revealed more percentage of daily time in OFF (from 4.2% to 13.6%), with akinesia and wearing-off, especially in the late afternoon (Figure 6.3). On the first day, false positive dyskinesia appeared when traveling by public transport.

Later, at 58 years of age, the patient was even more aware of morning akinesia and wearing-off symptoms and could quite accurately determine the duration of the OFF episodes. The number of levodopa intakes was increased

6.3 Improving Awareness of the First Motor Fluctuations

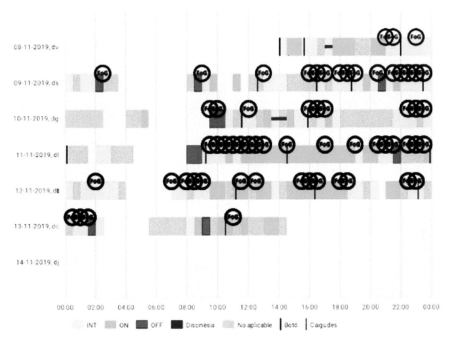

Figure 6.2 First STAT-ON™ report. It shows OFF and intermediate periods in the morning. Intermediate periods are also shown, especially in the afternoon (however, the patient denied having motor fluctuations).

from three to four, and the third STAT-ON™ sensor confirmed the accuracy of the patient quantifying the OFF episodes (Figure 6.4).

Discussion: In this clinical case, STAT-ON™ was very useful since it helped the patient and the neurologist detect early motor fluctuations. The sensor allowed the patient to understand better and to know his symptoms with good accuracy. The presented case is a long supervision period summary (around 3 years), showing different advantages when using this technology (noticing the neurologist the appearance of early, nonreported symptoms by the patient, helping the patient for a better knowledge of his disease and its evolution and promoting a good understanding between the patient and his doctor).

Conclusions and take-home messages: In conclusion, using the STAT-ON™ sensor can increase the awareness of motor fluctuations in patients with PD and help neurologists detect them earlier.

158 STAT-ON™

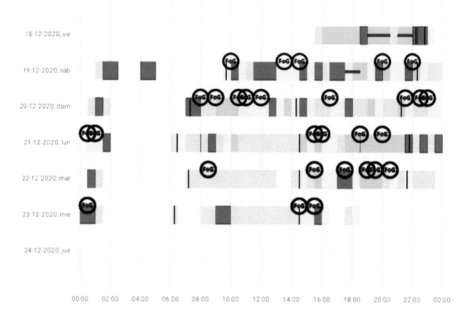

Figure 6.3 Second STAT-ON™ report. The patient is now aware of his morning akinesia and wearing-off episodes. An increase in OFF time and decreased FoG episodes occurred (compared to Figure 6.2).

6.4 Complimenting a Poor Patient's Interview about Her Motor Complications

Responsible professional: Dr. Tània Delgado
Hospital Parc Taulí. Sabadell (Barcelona)

It is about a patient with some problems identifying her motor symptoms correctly, making it very difficult to maintain the necessary interview at the doctor's office during the visit. The patient as the following:

Personal history: 61-year-old woman with an 11-year history of Parkinson's disease. No family history of neurological diseases, with the following personal history: hypertension, vitiligo, right knee prosthesis, and nephrolithiasis.

Parkinson's disease history: In 2011, she attended our clinic due to a recent onset left-hand tremor without other associated symptoms. At physical examination, she had left arm tremor at rest and difficulties with gait without left arm swinging. A cranial MRI was performed, showing a left posterior thalamic chronic microhemorrhage.

6.4 Complimenting a Poor Patient's Interview about Her Motor Complications

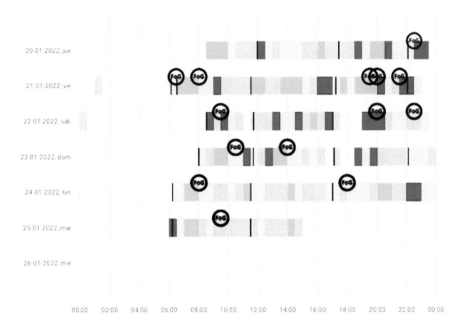

Figure 6.4 Third STAT-ON™ report. In this report, it is shown that FoG is less frequent. However, in this year, the patient complains of having more intense OFF periods.

Parkinson's disease (HY stage 1) was diagnosed, and treatment with rasagiline and pramipexole up to 1.05 mg/day were started, with moderate improvement in tremor.

In the following years, the tremor worsened, and a gait disorder appeared (left leg dragging). Levodopa was started up to 400 mg/day in September 2016 with a good clinical response. She also reported nonprogressive cognitive deficits. No visual hallucinations or delusional ideation were present. A neuropsychological study showed front-subcortical impairment according to Parkinson's disease. In November 2017, she began to have beneficial facial and extremities dyskinesias, and amantadine was added.

From November 2019, she noticed greater difficulties in carrying out her daily activities, and her tremor worsened. As a result, rasagiline was changed to safinamide without a clear benefit.

She did not come regularly for check-ups until January 2021. At that visit the patient reported being worse, but she could only explain that her "tremor" had increased. She used the term "tremor" to refer to tremor and dyskinesias, without being able to quantify it, nor to determine if there was a schedule, or there was a relationship with the medication intake.

At that time, the medication schedule was as follows: safinamide 100 mg (1-0-0), amantadine 100 mg (1-1-0), levodopa-benserazide 150 mg–150 mg-150 mg (at 8 h-15:30 h-23 h), and pramipexole 1.05 mg (1-0-0).

Physical examination: In ON state, she has normal facial expression, no speech problems, no tremor, clumsiness of the hands 2/2/1 bilateral, movements legs one bilateral, no rigidity, no bradykinesia.

It is observed moderate facial, axial and extremity dyskinesias, and dystonic posture in the arms during gait. Gait with normal steps, slightly unstable. Normal postural stability. An HY = 2 was determined, and a UPDRS-III of 13.

STAT-ON objective of use: In this case, it was suspected that the patient had both, dyskinesias and OFF periods. However, the information provided by the patient was very limited and misleading since she confused tremor and dyskinesias.

The objective of using the STAT-ON™ was to determine if there were motor fluctuations and if the called "tremor" episodes were related to OFF periods or dyskinesias, and, subsequently, to allow for optimizing the treatment.

Diagnosis and decision-making: STAT-ON™ was used for 6 days (with a total of 61 h of recording). The total OFF time, during this period, was 12.4 h (20%), an intermediate time of 9.7 h (16%), and an ON time of 17 h (28%) were obtained. The total time with dyskinesias was 4.8 h (8%) (see Figure 6.5 for details and distribution).

In general, she presented intermediate-OFF time from 1 to 5 pm and benefit dyskinesias at 10 am and 6 pm.

Discussion: The results showed that the patient presented OFF episodes at the midday levodopa intake. After the morning and midday levodopa intakes, beneficial dyskinesias were also detected.

Given these measurements, it was decided to lower the total levodopa dose and increase the frequency of doses to 100 mg in four doses (at 8-1-6-11 h). The patient improved clinically, reporting less dyskinesia and sustained response to treatment.

Conclusions and take-home messages: In this case, the STAT-ON™ device provided us with valuable information that could not have been obtained with the interview or the patient's diaries. As the device determines the presence and duration of the OFF periods, as well as dyskinesias, giving an objective perspective of the patient's daily motor status, this information permits the

6.5 Indirect Detection of Probable PD Nonmotor Fluctuations (NMF)

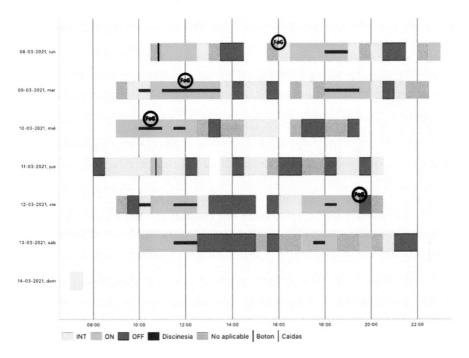

Figure 6.5 The STAT-ON™ report generally shows ON periods with dyskinesias in the morning, followed by OFF period between 1 and 5 pm. Around 6 pm, she experiences ON periods with dyskinesias again.

neurologist the proper adjustment of the treatment, obtaining a good clinical response.

As a final conclusion, it can be said that when the interview with the patients is poor or it is difficult for them to understand how to complete the patient's diaries, the STAT-ON™ device can be a very good alternative, helping the understanding of the patient's motor state.

6.5 Indirect Detection of Probable PD Nonmotor Fluctuations (NMF)

Responsible professional: Dr. Asunción Avila
Complex Hospitalari Moisès Broggi. Sant Joan Despí (Barcelona)

Parkinson's disease history: The patient is a woman. In July 2005, a 67-year-old female was referred to our movement disorders unit for the 1-year duration of nondisabling intermittent resting tremor in her right hand. Her past medical history revealed arterial hypertension.

On neurological examination, the patient had normal cognition. An intermittent mild resting tremor was observed in the right hand, as well as mild signs of cogwheel rigidity and bradykinesia. Gait and balance were normal and postural reflexes (UPDRS-III 11, Hoehn–Yahr 1.5). The diagnosis of idiopathic Parkinson's disease (PD) was entertained. During the first years, the patient had a good and maintained response to treatment with rasagiline 1 mg once daily and ropinirole extended release 8 mg once daily.

Three years after the diagnoses, the treatment was optimized with 300 mg/day of levodopa/benserazide with good clinical benefit. In July 2020, when she was 72, she came urgently to our Unit because she began experiencing an unpleasant feeling of emptiness in her abdomen that appeared in the afternoon, with no relation to the levodopa dose or other drugs. She described her uncomfortable sensation as *"nervousness,"* *"if something was missing,"* and *"if I was hungry."* The episodes could last between minutes and hours. The patient said that *"she didn't experience painful sensations."* In addition, she said that *"when it happened to her, she was useless, and she couldn't do anything."* Nevertheless, she denied having motor fluctuations (MF), freezing of gait episodes, or dyskinesias. Over the next 5 months, subsequent trials of modifying treatment were not helpful:

- Increasing the number of levodopa intakes and decreasing their intervals.

- Increasing the total dose daily of levodopa up to 800 mg/day with benserazide in four doses.

- Increasing daily dose of ropinirole extended-release up to 12 mg once daily.

- Adding opicapone 50 mg in a single daily dose.

- Adding entacapone 200 mg with some dose of levodopa.

The patient did not experience improvement with any treatment, and some modifications were abandoned in the first days because she felt worse. At the physical examination, she had mild right-sided parkinsonism with minimum right-hand tremor at rest and bradykinesia (UPDRS-III 5, Hoehn–Yahr 1.5). The psychiatrist of our movement disorders unit evaluated the patient without diagnosing her with any psychiatric pathology suggestive of specific treatment.

STAT-ON™ objective of use: Given the persistence of the symptoms, it was decided to use STAT-ON™ in order to try to identify slight motor fluctuations (MF) that the patient could not identify herself.

6.5 Indirect Detection of Probable PD Nonmotor Fluctuations (NMF) 163

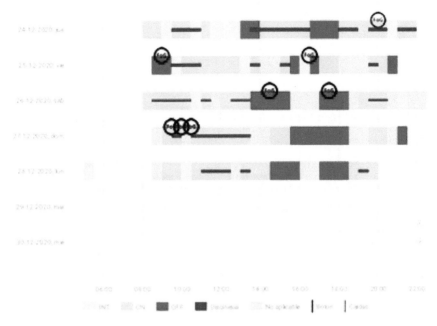

Figure 6.6 STAT-ON™ report showing a predominant OFF period between 2 and 7 pm, in coincidence with the experienced nonmotor fluctuation.

After the use period of 5 days, it was appreciated (Figures 6.6 and 6.7) that the patient presented OFF periods in the afternoon, especially between 2 and 7 pm, which coincided with the uncomfortable abdominal sensations that she had described. The Holter also recorded some isolated freezing of gait (FoG) episodes.

Diagnosis and decision-making: After the detection of the motor fluctuations (MF) that the patient was unable to detect herself, it was considered that, probably, the patient experienced nonmotor fluctuations (NMF) in the form of anxiety, abnormal abdominal sensations, and/or restlessness in coincidence with MF.

The patient was treated with rasagiline (1 mg/day), levodopa/benserazide (200/50 mg) in four doses daily, and ropinirole extended release 12 mg once daily. The patient was informed about the STAT-ON™ results and our suspicion about the coincidence between the MF and NMF periods. Additionally, we decided to substitute rasagiline for 100 mg of safinamide once a day.

After this intervention, the patient significantly improved her probable NMF. Three months later, a new STAT-ON™ registration was made during

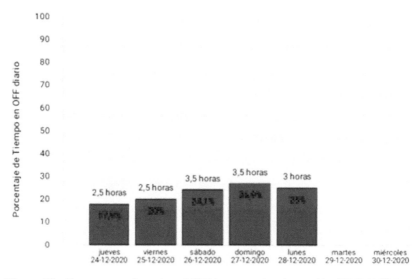

Figure 6.7 Percentage and number of OFF hours per day detected by STAT-ON™ (December 24 to 28, 2020).

3 days, and an evident reduction of the daily OFF time was detected, confirming the correlation between the patient's MF and NMF periods (see Figure 6.8).

Discussion: In Parkinson's disease, nonmotor symptoms can fluctuate (NMF) like motor symptoms (MF) [1]. An NMF can be seen as any change in the severity level of the nonmotor symptoms [1, 2], and many patients with MF also experience NMF with a prevalence in the range from 17% to 100% [2, 3].

This wide variability may be due to the difficulty in the identification of NMF since the diversity of fluctuating nonmotor symptoms, its largely subjective nature, and a frequent lack of perception of NMF despite the high impact of nonmotor symptoms on the autonomy and quality of life of the patient.

Twenty-eight percent of patients who experience both MF and NMF complain that NMF is more disabling than MF [3, 4]—psychiatric nonmotor symptoms related to motor symptoms in their timing and number of ON–OFF switches. The NMF can be present simultaneously with or later than MF [3].

Considering these PD aspects, it can be said that identifying FM using STAT-ON™ can help the neurologist indirectly identify the patients' NMF.

Conclusions and take-home messages:

- NMF include any nonmotor symptom change in the severity level, and their identification is difficult in clinical routine.

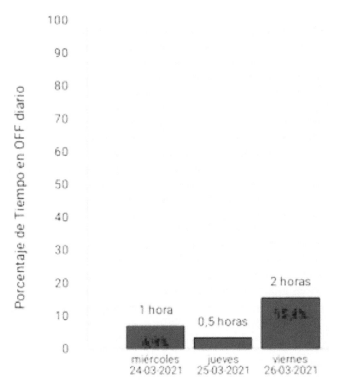

Figure 6.8 Percentage and number of OFF hours per day detected by STAT-ON™ 3 months later (from March 24 to 26, 2021).

- NMF develops simultaneously or after FM, so the identification of FM by STAT-ON™ can help professionals to identify NMF indirectly.

6.6 Deciphering the Patient's Complaints using STAT-ON™

Responsible professionals: Dr. Alexandra Pérez-Soriano and Dr. Núria Caballol
Unitat de Parkinson. Centro Médico TEKNON. Barcelona.

Personal history: A 68-year-old woman attended in 2020. She was diagnosed with PD in 2015. She graduated when she was 22 years old and worked in the bank sector until her retirement when she was 55 years old.

In her personal history, it was only remarkable dyslipidemia and hypothyroidism. Regarding family history, no other members had been diagnosed with PD. However, her mother, grandmother, and two uncles suffered from tremors.

Parkinson's disease history: Regarding premotor symptoms, hyposmia and a possible REM – Behavior Disorder (RBD) and constipation, were present. She had never been diagnosed with depression, although she defined herself "*as an obsessive and anxious person.*"

Her PD started with a rest tremor in her right leg and a rest and action tremor in her left arm and left leg. Initially, she was diagnosed with anxiety and received treatment with escitalopram and benzodiazepines for 3 months. Due to the persistency of the tremors, levodopa/carbidopa 300 mg was then started along with pramipexol-immediate release 0.18 mg with a good response. In 2016 she started to have motor fluctuations in the form of delayed-on in the morning, thus safinamide was added. From 2016 to 2020, her symptoms were quite well-controlled with levodopa-carbidopa 300 mg, pramipexol 1.05 mg, and safinamide 100 mg (levodopa equivalent dose of 655 mg/day).

In January 2020, she complained of bothersome movements in her right leg that appeared while standing. She referred to these right-leg movements as "*tremors,*" though during her follow-up visit, the movements clinically seemed dyskinesias. She also described that her right foot turned inwards, suggesting dystonic right foot movements associated with pain in her right leg. OFF periods with pain and right foot dystonia were suspected, and opicapone 50 mg was added in June 2020 to the treatment with partial relief.

However, in 2021 she noticed a worsening of her right leg movements that were especially annoying while standing and talking to someone in the street. Due to these symptoms, she avoided going out and stayed at home more than usual. Axial dyskinesia was also seen at the clinical examination hence amantadine 200 mg/day was added to her treatment schedule.

In December 2021, the equivalent levodopa dose was 855 mg/day, and her treatment was as follows: levodopa-carbidopa 100 mg 1-1-1 (300 mg/day), pramipexol 1.05 mg 1-0-0, safinamide 100 mg 1-0-0, opicapone 50 mg 1-0-0, and amantadine 100 mg 1-1-0.

In 2021, when she was asked about "OFF states," she was doubtful at first, saying that "*she wasn't sure she was having any*" but mentioned that she "*may be having 2 or 3 h a day in which she noticed worsening of dexterity in her left arm and leg, and generalized rigidity and slowness of movements.*"

Regarding the nonmotor symptoms, she complained of having anxiety and excessive sweating, but not in the OFF state. She was not sure about the amount of dyskinesia per day.

Physical examination: The progression of the UPDRS-III ON over the last years was from a UPDRS-III ON of 9 and an H and Y = 2 in 2019 to a UPDRS-III ON of 19 and a H and Y = 2, in 2021.

6.6 Deciphering the Patient's Complaints using STAT-ON™ 167

At the physical examination in the ON state, left rigidity and left bradykinesia were observed, along with axial and right leg dyskinesia.

In the OFF state, her left bradykinesia and rigidity worsened, and rest tremor was present in her right leg. Freezing of Gait was not observed.

STAT-ON™ objective of use: Although, in the present case, the presence and duration of the motor fluctuations could have been inferred from the clinical interview, the patient still showed many doubts regarding several aspects of her motor symptoms. Therefore, the objectives of using the sensor were:

- To confirm the existence of OFF periods.

- To explore more details about the timing and duration of the OFF periods.

- To confirm if the annoying right leg movements were dyskinesia.

- To analyze the timing and duration of the suspected dyskinesia.

The patient was instructed to wear the sensor for one week. She was trained to press the button at the time of the levodopa dose. At the same time, she was given a simple diary to write down her daily activities, as well as the precise moment in which her right leg movements became "annoying."

Diagnosis and decision-making: The report of the STAT-ON™ was analyzed with the patient. The dyskinesia and OFF periods shown by the report (Figure 6.9) were checked day-by-day along with the patient diary shown in Figure 6.11.

The diary is helpful if the patient collects some activities that could be false positives (going by car or transport, sweeping, etc.). Nonmotor symptoms (anxiety and pain) can also be recorded to check if the nonmotor fluctuation occurs along with the motor OFF state.

We can do a detailed analysis of Figure 6.9 contents:

- She came to our clinic to put the sensor on Friday 10. She arrived at her home at 10 am, and according to the report, she was in the ON state with dyskinesia, although she did not indicate the dyskinesia as bothersome that day in the diary (Figure 6.11). At midday and night, she was in OFF state. In total, 4 h in OFF were detected by the sensor on Friday 10 (Figure 6.9B), which represents 30.8% of her total wake time.

- On the second day (Saturday 11), the patient said she had morning akinesia for 20 minutes. Afterward, she was in an intermediate state, and right leg dyskinesia was detected. In the afternoon, the patient explained

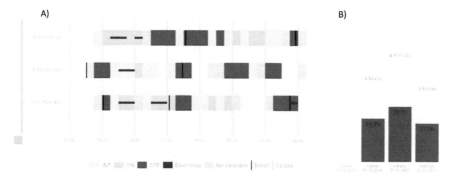

Figure 6.9 (A) Results of the first 3 days. After the morning akinesia period, she had ON with bothersome dyskinesia. (B) Total number of hours in the OFF time distributed per day.

that she was shopping, and although she was walking and active, she was in the OFF period.

- On the third day (Sunday 12), she marked 2 h in which her right leg movements were present and especially annoying: at 1 pm and 10 pm (Figure 6.11). This last dyskinesia period on Sunday 12, just after taking the levodopa dose (Figure 6.9A), could raise the possibility that her dyskinesia could be biphasic. The day before, she also had dyskinesia at 11 am while in the intermediate state.

Overall, in the 3 days analyzed, the OFF periods were present in the morning, midday, and night. The sensor also helped to show that the motor state was worse after midday and that she was better after the first morning levodopa dose, despite the bothersome dyskinesia.

In the report of the next 4 days, a better motor state was generally observed, with less OFF time per day (Figure 6.10):

- On the fourth day (Monday 13), she was doing housekeeping and reading in the morning. In her diary, she marked annoying symptoms at 10 am, 11 am, and 12 am, while doing these activities. (Figure 6.11). The report of the STAT-ON™ device showed a dyskinesia period from 10:15 am to 1:45 pm (Figure 6.10A).

- On the fifth and sixth days of monitoring, the patient marked bothersome dyskinesia on Tuesday 14 at 1 pm, while cooking and at 6 pm while she was shopping. Nonetheless, on that day, the STAT-ON™ did not capture the dyskinesia. Similarly, on Wednesday 15, she went out to a restaurant to have lunch at 2 pm, noticing the supposed annoying dyskinesia again, but the sensor did not detect them.

6.6 Deciphering the Patient's Complaints using STAT-ON™

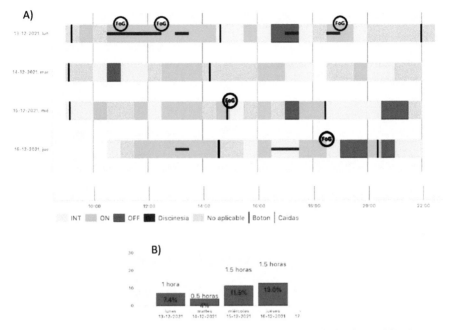

Figure 6.10 Compared with the previous 3-day monitoring period (Figure 6.9), the report shows a better motor state with an average of 1.5 h per day in OFF state.

After wearing the sensor, the patient's awareness of the motor complications improved and the clinical hypothesis of what was happening regarding motor complications was confirmed by the neurologist:

- OFF periods were confirmed. The clinician explained to the patient that during the morning, when she had morning akinesia, and she usually returned to bed to have more rest, she was actually in an OFF state. She also understood that OFF states related to dexterity worsening in her left extremities and slow movement, especially at midday and afternoon.

- Regarding dyskinesia, it was confirmed that they were bothersome most of the time and that they were both peak-dose and biphasic. She was instructed to call them correctly (dyskinesia and not "tremors").

Consequently, the treatment was adjusted. Levodopa total dose was reduced, and levodopa doses were fragmented into five doses: levodopa-carbidopa (five doses of 50–250 mg/day), pramipexole 1.05 mg (1-0-0), safinamide 100 mg (1-0-0), opicapone 50 mg (1-0-0), and amantadine 100 mg (1-1-0).

The dyskinesia improved during 6 months. However, the OFF periods intensity worsened in the next appointments, and an impulse control disorder

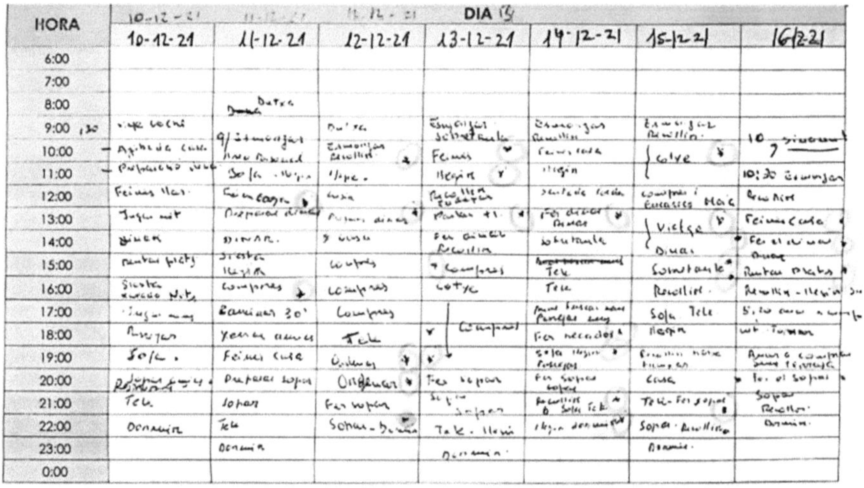

Figure 6.11 Patient's diary while using the sensor. In this case, the diary was useful for the patient to identify bothersome dyskinesia. The asterisks in green represented the moment the patient did not feel good, and the symptoms were bothersome to her. For example, the dyskinesia was annoying on Saturday 11, at midnight, or on Sunday 12, around 1 pm.

appeared (binge-eating). For these reasons, the patient is now being evaluated for deep brain stimulation (DBS).

Discussion: The identification of PD motor complications, in the daily-clinical practice, can be quite well recognized in a thorough clinical interview using the appropriate clinical scales and the Hauser diaries [5–7]. However, while for some patients, the motor symptoms and the motor fluctuations are easily recognized by using these methods, for some other patients, these symptoms cannot be easily identified.

In these cases, the clinical interview becomes more arduous, affecting the capacity of the neurologist to adjust treatment adequately. In addition, these patients are usually not deemed good candidates for clinical trials due to difficulties recognizing symptoms and completing Hauser's diaries. In this context, wearable sensors such STAT-ON™ can be useful to objectively assess motor fluctuations as well as to educate and empower the patient, improving self-awareness to detect motor symptoms and motor complications [8]. Moreover, in an outpatient setting, clinical scales and Hauser's diaries are not always used for reasons of time. Therefore, using the STAT-ON™ sensor can be a "novel way" to objectively assess the motor state of the patient during the day in a real-life scenario.

In the current case, the sensor detected motor complications that the neurologist had already suspected during the clinical interview. In general, the patient knew that sometimes she had OFF states affecting dexterity in her left extremities, and even though sometimes she named the right leg movements as "tremors," dyskinetic right leg movements were observed and suspected by the clinician. However, the use of the STAT-ON™ helped both the patient and the neurologist to increase the degree of self-awareness and better understand the main times when the symptoms became especially bothersome.

The sensitivity and specificity of the STAT-ON™ sensor for detecting dyskinesia are 95% and 93%, respectively, for strong or mild trunk dyskinesia [9]. The sensitivity is lower (39%) for mild limb dyskinesia. Despite these limitations, in our case, the STAT-ON™ report clearly showed the ON dyskinesia periods and only missed mild dyskinesia that the patient referred to when she was out shopping or eating. This indicates that the bothersome dyskinesia was not necessarily more intense but made them bothersome because they were present when she was doing activities in which she was in public or concentrating at home (reading and cooking).

Conclusions and take-home messages: In addition to the very good characteristics of STAT-ON™, identifying and analyzing the motor symptoms related to PD, it is convenient to emphasize that:

- It enables the precise identification of when a certain symptom (dyskinesia or OFF) is bothersome or disabling.

- It can be a very useful tool to educate the patient regarding suffering motor complications. For example, in the presented case, before wearing the sensor, the patient sometimes referred to the dyskinesias as "tremors" and after wearing the sensor, the patient recognized her dyskinesias better.

- For the neurologist, STAT-ON™ offers the possibility to better understand the patient's complaints and adjust the dopaminergic treatment.

6.7 Ambulatory Monitorization of a Patient with Advanced PD

Responsible professional: Dr. López-Ariztegui Núria
Movement Disorders Unit of the Hospital Universitario de Toledo. Toledo.

Personal history: The patient is a 70-year-old female, in follow-up at the movement disorders unit since she was 58 years old for PD and poor motor control in recent years. There is no remarkable personal history except for hypercholesterolemia, surgery for hallux valgus and saphenectomy, and family history of PD in several paternal relatives.

Parkinson's disease history: In 2010, at the age of 58, she consulted for several months of rest and postural tremor in the left arm, and loss of agility as well as hyposmia. She presented left asymmetric rigid-akinetic-tremor syndrome on examination with UPDRS-II = 3 and -III = 12.

The following complementary tests were carried out:

- DaTSCAN ioflupano (123I): hypo uptake of both putamens, asymmetric with greater right involvement.

- Cerebral MRI: within normality.

- Laboratory tests with biochemistry, complete blood count, copper, and ceruloplasmin: within normality.

- Genetic analysis: heterozygous variant gly2019Ser of the LRRK2 gene.

With the diagnosis of PD Hoehn and Yahr stage I, a treatment with rasagiline was started. After 6 months, the transdermal rotigotine was added up to 12 mg daily, with clinical improvement in tremor and daily life activities (DLA).

In 2013, she reported worsening of mobility with left leg dystonic and difficulties in DLA and sports such as swimming. Levodopa/carbidopa (LD/CD) was added up to 300 mg in three doses with significant motor improvement. She remained stable for 2 years, and in 2015, she started with morning akinesia and fragmentation insomnia due to nocturnal akinesia. An uncontrollable behavior compatible with impulse control disorder (ICD) in the form of kleptomania emerged. Since the ICD provoked marked anxiety, the dose of rotigotine was reduced to 8 mg, and the dose of LD was increased to four doses of 100 mg daily, with improvement in the ICD.

In the following years, she continued with motor fluctuations, mainly nocturnal and morning akinesia and mild wearing-off, and she maintained independence in DLA, although she had to give up swimming.

She referred mild and nondisruptive choreic dyskinesias. Different treatment adjustments were made, fragmenting LD/CD, adding opicapone, and an attempt to change from rasagiline to safinamide, but she did not tolerate it due to adverse events (AE). Lastly, second-line therapies (SLT) were proposed, but patient and relatives were reluctant to do all of them since, with medication adjustments, she felt relatively well.

From the very beginning, she had been diagnosed with anxious depressive symptoms controlled by psychiatry with escitalopram and bromazepam.

Over time, nonmotor symptoms (NMS) appeared: low back pain related to lumboarthrosis that worsened in the afternoon, hypertension in off episodes that require various antihypertensive medication adjustments, episodes of excessive sweating, nocturia, constipation, mood, and fluctuating sleep with the need for antidepressant treatment adjustments.

In evaluation after the COVID pandemic, the patient was worse; she continued the treatment with rasagiline 1 mg, rotigotine transdermal 8 mg LD/CD 800 mg in five daily doses, and opicapone 50 at night (LEDD 1540 mg): she spent the afternoon sitting without activity because she was tired with lower back pain.

She had morning akinesia lasting an hour and disruptive wearing-off around 1 pm that interfered with her DLA. Since she was still reluctant to SLT, ambulatory monitoring with STAT-ON™ was scheduled to characterize the OFF episodes.

Physical examination: MDS-UPDRS-III-Off state = 42. It was observed left asymmetric rigid-akinetic tremor syndrome with slow but autonomous gait and freezing of gait (FoG) at the beginning of walking and when turning.

MDS-UPDRS-III-On state = 9, with mild hypophonia, postural alteration with Pisa to the left, minimal asymmetry in bilateral tapping maneuvers with mild axial choreic dyskinesias and left leg, without tremor.

MDS-UPDRS-I 0 = 9 (mood and anxiety, fatigue, sleep, pain, constipation, urinary), II = 10, IV = 6. The Hoehn and Yahr state is 2 in ON, 3 in OFF state.

STAT-ON™ objective of use: The patient was diagnosed with familial PD concerning heterozygous mutation gly2019Ser of the LRRK2 gene. A complicated PD stage is observed with motor and nonmotor fluctuations and dyskinesias, susceptible to SLT.

Although the presence of disruptive motor fluctuations was clear from history, the patient and family underestimated them. They related them with fatigue and did not decide on any SLT solution. It was proposed to carry out ambulatory monitoring with STAT-ON™ to quantify and characterize her OFF moments and achieve adjustment and adherence to new therapeutic measures. The patient was instructed to press the event button with each LD/CD intake.

Diagnosis and decision-making: The STAT-ON™ report covered 6 days (Figures 6.12 and 6.13) and showed an active patient who walked an average of 18,000 steps per day, but she was less than 50% of the monitored time in

Días monitorizados:	6
Tiempo monitorizado:	83.5 horas

N° episodios FoG:	63
Media episodios FoG/día:	10.5±8.4
Media minutos andando/día:	183.8±67.4
Media número de pasos/día:	18041.5±6699.3
Tiempo total sin diagnóstico (% tiempo monitorizado):	6 horas (7.2%)
Tiempo total en OFF (% tiempo monitorizado):	17.5 horas (21.0%)
Tiempo total en Intermedio (% tiempo monitorizado):	20.5 horas (24.6%)
Tiempo total en ON (% tiempo monitorizado):	39.5 horas (47.3%)
Tiempo total con discinesias (% tiempo monitorizado):	11 horas (13.2%)
Umbral personalizado de bradicinesia (Fluidez de zancada) >8 óptimo <6 subóptimo	6.6±0.5

Figure 6.12 Summary of STAT-ON™ report. As indicated, it was detected a 21% of OFF Time, a 47.3% of ON time, and a 24.6 % of intermediate state. She was suffering from dyskinesias during a 13.2% of the monitoring time.

ON state. She spent 21% of the time in the OFF state and 24.6% in the intermediate state with the following distribution:

- Morning akinesia, with episodes of FoG, and a latency in the effect of the first dose of more than 30 minutes.

- Episodes of wearing-off in the rest of the LD/CD doses, with a variable duration of 30–120 minutes with a transition between intermediate and OFF state, associated with FoG phenomena less frequently than in the morning.

- Dyskinesias generally appeared in the afternoon but accounted for only 13% of the monitored time, and the patient reported them as not disruptive.

After discussing these results with the patient, it was decided to perform an apomorphine test, which was positive at the dose of 4 mg without adverse effects. Treatment was adjusted by increasing the dose of LD in the first and second levodopa intakes up to 200 mg. Subcutaneous apomorphine injections were started for OFF periods rescue. In the subsequent visit, the patient explained that she recognized well the OFF episodes, and that she only used the apomorphine midday rescue, if she had to go out to do some activity.

6.7 Ambulatory Monitorization of a Patient with Advanced PD

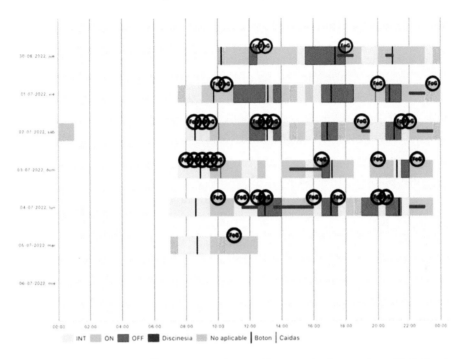

Figure 6.13 The STAT-ON™ report shows intermediate state in the morning with FoG episodes. The report demonstrates OFF states associated with the LD/CD doses. Dyskinesias were especially present in the afternoon.

Discussion: This case reflects a common problem found in the clinical practice with patients with PD: the difficulty for many patients and their relatives to recognize the OFF symptoms [1]. Recognizing OFF time can be especially difficult when nonmotor symptoms such as fatigue, pain, or mood disorders dominate the OFF periods. In these cases, when performing ambulatory monitoring and reviewing it with the patient, they can realize that their symptoms are related to LD doses and help them to have greater therapeutic adherence.

A second common problem, in clinical practice, is that the patient and their relatives do not understand the need to change the therapeutic strategy with the transition from a conventional one to an SLT [2]. Sharing with the patient the STAT-ON™ report and showing them in the registry the changes that occur during the day, can help in deciding to move to an SLT or a device-aided therapy.

Conclusions and take-home messages: The use of devices for ambulatory monitoring of PD patients, such as STAT-ON™, helps the physician to know the real motor state of these patients in their daily life. Moreover, it is of great help for the patient to realize why they do not feel well at a precise moment of the day and, to establish a relationship with medication. Thus, they can understand the different therapeutic decisions that must be made to improve their clinical situation and quality of life, such as the initiation of the SLT.

6.8 Improvement of the Patient's Awareness of the Advanced PD Stage and the Need for a Second-line Treatment

Responsible professional: Dr. Sònia Escalante
Hospital Verge de la Cinta. Tortosa (Tarragona).

Personal history: A 73 years old female with a 15-years history of Parkinson's disease. Some additional data are: treated hepatitis C virus infection, with undetectable viral load, depression and anxiety. Previous surgeries: appendectomy, bilateral knee prosthesis, hallux valgus. No family history of Parkinson's disease.

Parkinson's disease history: When she was 58, she was diagnosed with PD. Her initial symptoms were left arm rest tremor and bradykinesia with good response to pramipexole and rasagiline. She had no balance impairment, constipation, or smell loss. She also was diagnosed with sleep disorder, suggestive of REM sleep behavior disorder (RSBD), with good response to clonazepam.

She had a brain MRI that showed leukoaraiosis and a DaTSCAN with a significant reduction of right putamen's dopaminergic activity.

Five years after the diagnosis, the tremor was bilateral, and she started with visual hallucinations and eating behavior disorder. This was controlled by reducing the dosage of dopaminergic agonist agents and starting low doses of levodopa.

When she was 69, she developed motor fluctuations (morning akinesia and wearing-off), needing an extra dose of levodopa. Meanwhile, she showed symptoms of cognitive decline.

At the age of 71, she had morning akinesia lasting 1 h, and nondisrupting dyskinesia appeared. She needed to take 5 levodopa doses, but some were not effective.

At this point, we discussed with the patient the second-line treatment options. She did not meet the DBS criteria because of her cognitive impairment, and apomorphine was not considered a suitable option because of her eating behavior disorder. We explained to her the intestinal levodopa infusion therapy, but the patient was scared and claimed that *"she had good days and*

6.8 Improvement of the Patient's Awareness   177

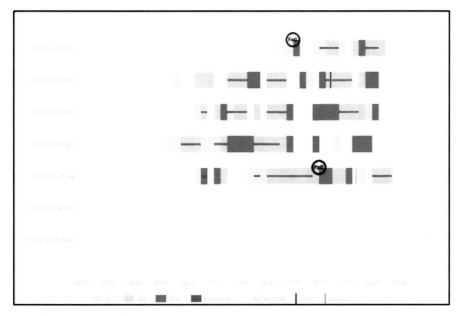

Figure 6.14 STAT-ON™ summary report showing the above-mentioned situation.

bad days." Hauser's diary was not helpful because she checked 2 items at the same time (OFF and dyskinesias, for example) multiple times.

At that time, the treatment schedule was: rasagiline 1 mg, pramipexole 1.05 mg, levodopa/carbidopa 150 mg five times daily, clonazepam 0.5 mg at night.

Physical examination: In OFF: H and Y = 2.5, UPDRS-III: 29. Occasional Freezing. She needed help with some daily life activities. In ON: H and Y = 2, UPDRS-III: 12. Nondisrupting dyskinesia. Independent for all daily life activities.

STAT-ON objective of use: It was decided to use the STAT-ON™ Holter to record objective data and show them to the patient to make her aware of her real situation. The second foreseen objective was to see if this could help her to decide about a second-line treatment.

Diagnosis and decision-making: The STAT-ON™ device was used for 5 days with a total of 66 h of recordings (see Figures 6.14 and 6.15) with the following conclusions:

- Motor fluctuations were detected: wearing-off, morning akinesia, and nondisrupting dyskinesia.

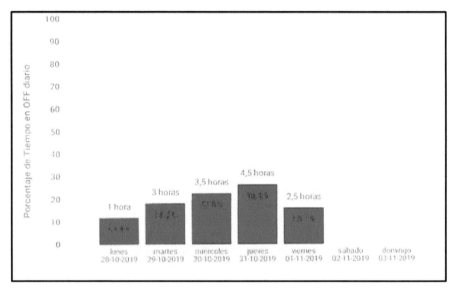

Figure 6.15 Percentage of daily OFF time.

- The total OFF time was 6.9 h (10.4%), the intermediate time was 5.1 h (7.7%), and the ON time was 16.3 h (24.7%).
- The total time with dyskinesia was 9.7 h (14.7%).
- No FoG episodes were detected during the ON periods, and she had dyskinesia almost all the time during ON periods.
- Nearly every day, she had more than 2.5 h of OFF periods, arriving, some days, to 4.5 h.

Discussion: Results obtained from the use of STAT-ON™ were very useful to show to the patient how complicated it was to manage her disease with only oral medications. She was aware of the presence of dyskinesia during almost all the duration of the ON periods and this limited further up-titration of levodopa.

This helped her to understand how frequent her OFF periods were and the difficulty of controlling her disease with only an oral medication approach. She is now treated with levodopa-carbidopa intestinal gel infusion, and the motor fluctuations are now much more well-controlled.

Conclusions and take-home messages: STAT-ON™ device provides objective information, which is extremely useful to optimize dopaminergic treatment, mainly when the information provided by the patient is not clear enough or when the neurologist suspects that the patients could be minimizing their symptoms,

In patients with advanced-stage PD, who might have problems in identifying ON–OFF periods, this device could be a valuable tool to detect candidates to a second-line treatment.

6.9 Identification of CANDIDATES to a Device-aided Therapy

Responsible professional: Dr. Diego Santos-García
CHUAC – Complejo Hospitalario Universitario de A Coruña. A Coruña.

Personal history: A 58-year-old right-handed woman was referred for PD evaluation. She presented:

- Idiopathic PD with the onset of symptoms 6 years before the visit.
- Previous treatment for pulmonary tuberculosis.
- Plaque psoriasis and psoriatic onychopathy are in remission.
- No known drug allergies.
- No cardiovascular risk factors.
- No toxic habits.
- Chronic constipation.

The treatment was: Sinemet Plus® 1-1-1-1-1 (at 8:45-11:45-14:45-17:45-20:45), Sinemet Retard® (1 pill at night), Ongentys®, Rivotril® (0.5 mg at night), Metoject®, folic acid. Daily dose of levodopa 700 mg/day. Levodopa equivalent daily dose (LEDD) is 975 mg/day. The patient had been treated before with safinamide and dopamine agonists (rotigotine and pramipexol) with no tolerability.

She was retired, living with her sister (principal caregiver), and having good family support. Concerning the familiar history: no cases of PD or any other neurological condition in her family.

Parkinson's disease history: In 2014, the patient started with a resting tremor in her right leg. She received some drugs without good tolerability: Artane® (dry mouth); Neupro® (nausea and vomiting); Mirapexin® (dizziness and constipation). In August 2018, L-dopa was increased, up to 400 mg/day. In October 2020, she started with motor fluctuations, and Xadago® (50 mg/day) was added to levodopa, but it was withdrawn due to dizziness and psoriasis outbreak. After this, she started with entacapone, and some months after this, it was changed to opicapone with very slight benefit.

The patient was referred to the CHUAC – movement disorders unit for evaluation and consideration about a possible device-aided therapy.

At that moment of evaluation (November 2020), the patient presented with predictable (morning akinesia; wearing-off) and unpredictable motor fluctuations (delayed-ON; no-ON; partial-ON) as well as dyskinesia, sometimes disabling for the patient (especially in mouth when she was in public with other people). During the OFF episodes, the patient developed tremor, rigidity, bradykinesia, anxiety, and sometimes fatigue with a bad mood.

A variability was observed depending on the days, with some days with fewer than 2 h of OFF time during the waking day and others with more than 4 h. In some moments, she felt fatigue and a worse mood with lack of motivation but without being especially worse in her movements.

Complementary tests:

- Cranial and cervical MRI (2018): without significant alterations.

- DaTSCAN (2018): bilateral striatal dopaminergic denervation with left side predominance.

Diagnosis:

- Parkinson's disease of about 6 years of disease duration (from symptoms onset).

- Motor fluctuations and disabling dyskinesia. Very good response to levodopa (Hoehn and Yahr = 2 and UPDRS-III 11 during the ON state).

- Minor depression, mild anxiety, constipation, urinary symptoms, fatigue, and REM sleep disorder as the most relevant NMS. Nonmotor fluctuations (fatigue, mood, and motivation).

Physical examination: The general and neurological examination was done without alterations

Motor assessment:

- UPDRS-IV: 8. OFF time 26-50%. Dyskinesia 1-25%.

- FoG-Q: 5. No significant freezing of gait (FoG) episodes.

- UPDRS-III-OFF (9:10): 35. Language 0. Hypomimia 1. Tremor 6 (0103110). Rigidity 4 (01111). Bradykinesia 16 (32212132). Posture 1. Gait 2. Postural reflexes 1. Global bradykinesia 4. Hoehn and Yahr 2.5.

- UPDRS-III-ON (10:20): 11. Language 0. Hypomimia 0. Tremor 1. Rigidity 1. Bradykinesia 9. Hoehn and Yahr 2.

Nonmotor assessment:

- PD-CRS: 100 (fronto-subcortical 71, cortical-posterior 29).
- NMSS: 47/360 (cardiovascular 0/24; sleep/fatigue 9/48; mood/apathy 7/72; perceptual problems 0/72; attention/memory 1/36; gastrointestinal tract 9/36; urinary symptoms 8/36; sexual dysfunction 1/24; pain and miscellaneous 12/48).
- BDI-II: 8/63. Positive for minor depression.
- QUIP-RS: 0. No impulse control disorder.
- PDSS: 144/150.

Quality of life and autonomy for activities of daily living:

- PDQ-39SI: 32/156 (mobility 19/40; activities of daily living 4/24; emotional well-being 5/24; stigma 0/16; social support 0/12; cognition 1/16; communication 0/12; bodily discomfort 3/12).
- PQ-10: 6/10.
- EUROHIS-QOL8: 27/40.
- ADLS: 60% (OFF); 90% (ON).

STAT-ON objective of use: The patient rejected the option of starting with device-aided therapy. Since the patient had not previously tolerated treatment with dopamine agonists and she did not want to consider deep brain stimulation, a levodopa/carbidopa infusion was proposed. Still, the patient refused the levodopa/carbidopa infusion therapy. Previously, many levodopa adjustments had been conducted without good response, partly due to the development of dyskinesia. Therefore, amantadine was added to her treatment. However, after some months with amantadine the patient stopped due to no tolerability (she felt impairment in movements).

To know the daily OFF time during the waking day, the STAT-ON™ Holter was used for 1 week in May 2021.

Diagnosis and decision-making: A clear pattern was detected from the report generated, and two different OFF periods throughout the day were identified. Firstly, in the morning (from 8 to 10 am) and then after midday (from 2 to 4 pm) (see Figure 6.16).

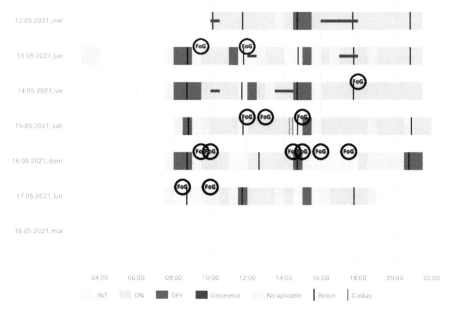

Figure 6.16 STAT-ON™ report summary showing well-defined OFF time periods in the morning and after midday. Many FoG episodes were also detected.

Moreover, a third moment of OFF time, though shorter, was detected around 12. PM. However, the status of the patient during the evening, in general, was better, with quiet ON time.

The patient was ordered to press the button at the time of taking levodopa, to find out the relationship between the episodes with these moments. The information collected was consistent with morning akinesia and wearing-off at the second and at third doses of levodopa during the day. Fewer time with dyskinesia was also detected. Interestingly, 14 FoG episodes were also captured. Only very few of the FoG episodes were during the ON time, whereas the rest of the FoG episodes were during the OFF time or intermediate state.

Although previously, the patient was asked about FoG episodes and answered that she did not having this symptom, after checking the monitoring records, she commented that very brief minor episodes could have appeared sometimes when she felt worse.

The patient's perception of presenting more FoG episodes as per what was recorded by the STAT-ON™, especially on Sunday 16 (Figure 6.17).

Daily OFF time ranged from 1 to 3 h (see Figure 6.18). After reviewing the monitoring record with the patient, she perfectly saw the presence of fluctuations throughout the day, and she agreed to start with levodopa/carbidopa

6.9 Identification of CANDIDATES to a Device-aided Therapy 183

Figure 6.17 Weekly FoG report showing that the worst day regarding the number of FoG episodes was on Monday 16.

infusion therapy. The patient was treated with Duodopa® in May 2021, with a very good response and tolerability.

Discussion: The present case is one example of the possible utility of the STAT-ON™ as a tool to identify a patient with advanced PD as a candidate for device-aided therapy.

This patient was a 5-2-1 criteria positive patient [12] with motor fluctuations, nonmotor fluctuations, and dyskinesia. According to the CDEPA criteria, this patient was an advanced PD patient [13]. A recent publication about the opinion of expert neurologists on PD using the STAT-ON™ showed that the STAT-ON™ could be a useful tool to detect advanced PD [14].

Interestingly, the monitoring record about the patient's state during the waking day obtained with the STAT-ON™ was useful for informing the patient and changing the decision about the therapy. Although she rejected initially to start with a device-aided therapy, after reviewing the STAT-ON™ results, she agreed to start with levodopa/carbidopa infusion therapy. Moreover, the STAT-ON™ made it possible to identify the presence of FoG episodes, which have not been previously detected with the clinical evaluation. The STAT-ON™ has also been validated with advanced-stage PD patients with levodopa-carbidopa intestinal gel or deep brain stimulation [15, 16]. Although it was not done, the STAT-ON™ could have been used for monitoring this patient's response to Duodopa® in this patient both in the short and long term.

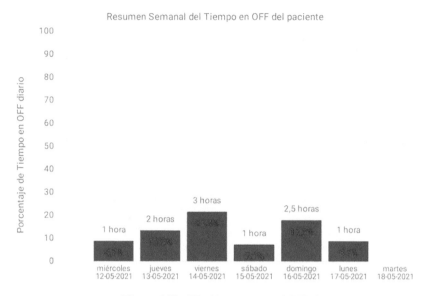

Figure 6.18 Weekly summary of OFF time.

Conclusions and take-home messages:

- STAT-ON™ was used in a patient with advanced PD with motor fluctuations and dyskinesia who rejected to start with a device-aided therapy.

- The type of motor fluctuations and OFF time was perfectly identified. FoG episodes were detected when they had not been previously identified with the clinical assessment.

- The information collected with the STAT-ON™ was useful for objectively showing the patient some complications of her disease (OFF episodes, dyskinesia, and FoG) and convincing her of the decision to start with a device-aided therapy.

- A correlation between motor OFF episodes and some NMS (fatigue and bad mood) was collected.

Finally, the patient was treated with levodopa/carbidopa infusion, and her symptoms improved.

6.10 STAT-ON™ Use for LCIG Tube Adjustment

Responsible professional: Dr. Jaime Herreros Rodriguez
Hospital Universitario Infanta Leonor (Neurology Department). Madrid.

Personal history: 73-year-old male with an excellent physical condition and cognitively intact. His personal history was unremarkable except for arterial hypertension treated with enalapril/hydrochlorothiazide (20/12.5 mg).

Parkinson's disease history: He was followed in the clinic for advanced Parkinson's disease (PD) (stage III of H–Y) with motor complications (wearing-off, no-on, and occasional ON–OFF phenomena). He was diagnosed with PD at the age of 65, and a good response to ropinirole (started in 2014) was obtained. After 2 years, levodopa was added with good clinical benefits. Opicapone was added in 2019, and safinamide in 2020.

    Treatment: safinamide (100 mg at dinner), opicapone (50 mg at breakfast), Levodopa/carbidopa 250 mg 1.5 (8 am)-1.5 (1.30 pm)-1/2(6 pm)-1/2 (9 pm), ropinirole extended-release (8 mg at breakfast).

    Due to poor medical control despite oral pharmacological optimization, it was decided to start with levodopa-carbidopa intestinal gel in continuous infusion (LCIG) in October 2021 (Duodopa: morning dose (18 mL); continuous dose (3.2 mL/h); extra dose (2 mL)).

    After 2 weeks of starting the LCIG, the patient reported a significant improvement in his daily activities and motor state. The motor state was recorded using the STAT-ON™ device.

    However, 2 weeks later, the patient-reported acute motor impairment without achieving a good clinical motor situation up to 6 h after starting the daily infusion of levodopa.

    Some decisions were made: the morning dose of LCGI was increased to 22 mL, an abdominal X-ray was requested to check the position of the LCIG tube (Figure 6.19, left), and it was prescribed a new use of the STAT-ON™ device to analyze the patient's motor state along the day.

STAT-ON™ objective of use: Considering the characteristics of the STAT-ON™ device, it was recommended its use in the present case to demonstrate the clinical worsening 2 weeks after LGCI was instituted (LCIG treatment with and without normal functioning). Results in Table 6.1 show the opinion and feeling of the patient at the beginning (2 weeks after the LCGI adjustment) and when the mispositioning of the tube was detected.

Diagnosis and decision-making: After confirming the LCIG tube mispositioning in the stomach, starting with prokinetics (domperidone three times daily) was decided. The patient recovered his good previous clinical situation (functional and motor) five days later. An abdominal X-ray control was done, showing the right positioning of the internal probe (Figure 6.19, right).

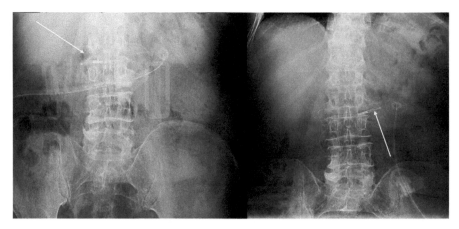

Figure 6.19 LCIG tube mispositioning in stomach (left) and right positioning in ileum (right). Yellow arrow points to the tail of the internal probe.

Table 6.1 Motor's state according to the patient opinion.

	Motor fluctuations (according to the patient's opinion)	
	2 weeks after LCIG tube adjustment establishment	LCIG tube mispositioning
Morning delayed-on	30 minutes	180 minutes (suboptimal)
"No-on"	Absent	Daily
Wearing-off	20% waking time	50% waking time
Extra dose	Usually not	4–5 times daily
ON–OFF phenomena	Absent	Occasionally

A new measurement period was scheduled with STAT-ON™ in order to objectively verify the worsening state described by the patient. In total, two STAT-ON™ reports were obtained (see the summary in Table 6.2):

- A report 2 weeks after LCIG tube establishment (improvement reported by the patient)

- Four weeks after LCIG tube establishment: patient-reported clinical worsening (LCIG tube mispositioning was detected).

The patient switched from one freezing of gait episode to 4.6 episode per day and walked 124 minutes less each day (26% less than before). The patient also walked 1392 less steps per day on average. A total of 58.4% of inactivity time was reported after probe mispositioning, against 43.2% prior to that change. With the tube right positioned, both OFF and ON times slightly increased due to the patient's activity.

Table 6.2 Summary of the STAT-ON™ report before and after the LCIG tube mispositioning.

	LCIG tube right positioning	LCIG tube mispositioning
FoG episodes	5	41
Fog average per day	1	4.6
Minutes walked per day	59	46.6
Average number of steps	6547.9	5155.3
Total time inactive	43.2	58.4
% Time in OFF	14.8	11.2
% Time intermediate state	18.8	10.9
% Time in ON	23.3	19.5
% Time with dyskinesia	10.8	10.9
Days monitored	13	7
Hours monitored	164.5	88

Discussion: LGIC is a second-line therapy that benefits selected PD patients' quality of life [17, 18]. However, therapy management and supervision are complex, and evaluating the therapy results frequently relies on the patient's opinion.

In the considered case, the etiology for the patient's worsening was a spontaneous wrong placement of the duodenal tube.

STAT-ON™ was useful to objectively quantify the clinical motor situation of PD patient. In this case, the neurologist was able to test motor worsening related to LCIG tube mispositioning compared with the previous clinical situation.

Conclusions and take-home messages:

- STAT-ON™ has been useful to show and quantify the patient's motor improvement or worsening due to the LCIG therapy.

- The monitoring was done in a home environment, and the information obtained was more precise than the clinical features detailed by the patient and his relatives.

- This enormous amount of information could be able to establish certain clinical patterns that point to one cause or another of LGIC dysfunction, in our case, bad placement of the internal probe.

6.11 Monitoring FoG and Second-line Treatment

Responsible professional: Dr. Iria Cabo López
CHUP – Complexo Hospitalario Universitario de Pontevedra. Pontevedra.

Personal history: A 69 year-old Spanish male patient, with a history of ischemic heart disease and dyslipidemia. He was an ex-smoker and had a moderate enolic habit. He was currently taking bisoprolol 5 mg OD, simvastatin 20 mg OD, ranitidine 300 mg OD, tamsulosin 0.4 mg OD, clopidogrel 75 mg OD, and olmesartan 20 mg OD.

Parkinson's disease history: The onset of his Parkinsonian symptoms started in March 2014 with "internal" tremor, slower right movements, and difficulty for walking since about 2 years. A clinical neurological examination revealed mild facial hypomimia, right arm, and leg rigidity ¼, right bradykinesia ¾, and left bradykinesia 2/4. Rest tremor was not present. However, his gait was slow with short steps. His feet stuck to the floor, and his right arm swinging decreased. At the time of diagnosis, UPDRS-II was 1, UPDRS-III was 11, UPDRS-IV was 0, and H and Y = 2.

In summary, the patient presented an akinetic-rigid syndrome with right dominant motor symptoms suggestive of idiopathic Parkinson´s disease. Laboratory test and cerebral magnetic resonance imaging were normal, and treatment with rasagiline and ropinirole was started.

In September 2014, he recognized a deterioration in his Parkinsonian symptoms with a worsening in his motor symptoms, and the dose of ropinirole was increased. In January 2015, levodopa/carbidopa immediate release was started because of motor impairment. There was an important improvement of rigidity and bradykinesia, though freezing of gait (FoG) episodes remained, especially when turning or in narrow places. He remained unchanged until March 2017 when he started with peak-dose dyskinesia and mild wearing-off. He also developed an impulse control disorder (hypersexuality) and a gradual withdrawal of ropinirole was required. In May 2018, axial symptoms became more evident with FoG episodes in his OFF's periods as well as levodopa-induced ON FoG.

Physical examination: In 2019, when the patient was evaluated with the STAT-ON™ for the first time, before initiating treatment with apomorphine, he was taking safinamide 100 mg OD, opicapone 50 mg OD and immediate release carbidopa/levodopa 100 mg five times per day.

Clinical neurological examination revealed left and right arm and leg rigidity ¼, right bradykinesia ¾ and left bradykinesia 2/4, no rest tremor, very slow walking with very reduced step length, his feet stuck to the floor, defragmentation of turns and decreased right arm swing, with several FoG episodes during the physical exam.

OFF: UPDRS-II: 3. UPDRS-III: 34. H and Y 2.5.

- ON: UPDRS-II: 1. UPDRS-III: 16. H and Y 2.
- UPDRS-IV: 3. NMSS: 50. PDSS: 132. FoG-Q: 13.
- Schwab and England: 80%. PDQ-39: 20. WHOQOL-8: 32.

STAT-ON™ objective of use: In December 2019, this patient was evaluated with the STAT-ON™ with the purpose of quantifying OFF/ON time and for the assessment of OFF/ON FoG episodes.

In October 2020, treatment with continuous infusion of apomorphine was initiated, and initial dose was adjusted according to the clinical response until motor control was achieved (infusion rate of apomorphine: 1.05 mL/h).

In December 2021, the patient was evaluated with the STAT-ON™ again, with the purpose of monitoring apomorphine response and assessing changes in FoG episodes.

Diagnosis and decision-making: When the patient was evaluated with STAT-ON™ for the second time in 2021, after initiating apomorphine, his clinical neurological examination revealed right rigidity ¼, left rigidity 0/4, right bradykinesia ¼, left bradykinesia 0/4. His gait improved with better step length and less defragmentation of turns.

- OFF: UPDRS-II: 7. UPDRS-III: 37. H and Y 2.5.
- ON: UPDRS-II: 3. UPDRS-III: 20. H and Y 2.
- UPDRS-IV: 5. NMSS: 30. PDSS: 136. FoG-Q: 10.
- Schwab and England: 90%. PDQ-39: 13. WHOQOL-8: 32.

The first STAT-ON™ report supported the diagnosis of advanced Parkinson´s disease with OFF and ON FoG episodes. The report showed a substantial number of FoG episodes (482) during the 4 days of registration, with an average of 96 episodes of daily FoG. FoG episodes were present in both ON and OFF periods.

Furthermore, average of daily OFF time was 24% while ON time was 21%. The average of OFF hours per day was between 3 and 5.5 h. Therefore, the STAT-ON™ provided confirmation of advanced PD report with more than 3 OFF hours a day, as well as a big number of FoG episodes, which led to instauration of a second-line treatment (Apomorphine infusion therapy).

The second STAT-ON™ report, in 2021, supported an improvement in his motor symptoms and more specifically, in OFF/ON number of FoG episodes. The report showed a substantial reduction in the number of FoG episodes (30) during the 4 days of registration, with an average of 6 episodes

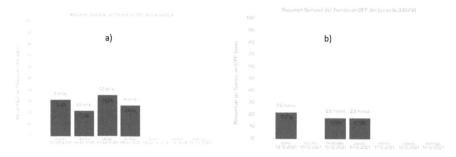

Figure 6.20 Percentage of daily OFF time (A) preapomorphine and (B) postapomorphine.

of daily FoG. Furthermore, average daily OFF time was 13%, while ON time was 46%. The average of daily OFF time was between 0 and 2.5 h a day.

Therefore, the STAT-ON™ reported a global improvement, specifically in OFF and ON time, as well as a reduction of FoG episodes after the instauration of Apomorphine infusion treatment.

Figures 6.20–6.22 show the difference observed between the pre and postapomorphine situations.

Discussion: Certainly, in this case, the device has been very useful to assess the motor fluctuations and the ON FoG episodes, but also to assess the motor state after the onset of a second-line therapy (use of apomorphine).

Conclusions and take-home messages: STAT-ON™ is very useful for completing information provided by the patient or the Hauser diary, providing accurate information about the motor state (ON, OFF, and FoG episodes). It is very useful for monitoring the effects of a second-line treatment.

6.12 Improving Motor Fluctuations with Variable Flow of Apomorphine Subcutaneous Infusion: The Role of STAT-ON™

Responsible professional: Dr. Jorge Hernandez-Vara
Neurology Department and Neurodegenerative Diseases Research group of the Vall d'Hebron University Campus. Barcelona.

Parkinson's disease history: A 73-year-old man was diagnosed with Parkinson's disease at 58. When he was 69, he experienced motor complications (motor fluctuations and dyskinesias) and was initially managed with oral antiparkinsonian drugs.

At the age of 71, motor fluctuations became refractory to conventional oral medication. At this moment, he was treated with levodopa/carbidopa

6.12 Improving Motor Fluctuations

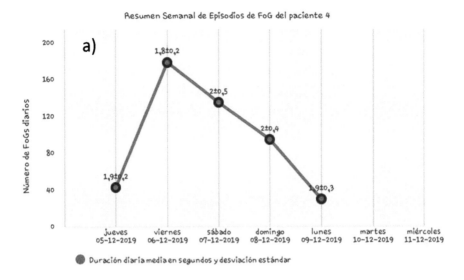

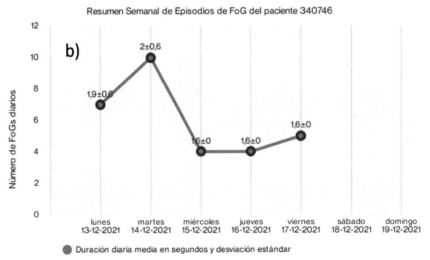

Figure 6.21 Number of FoG episodes per day (dot indicates the average duration) (A) preapomorphine and (B) postapomorphine.

immediate release (100/24 mg) six times per day, levodopa/carbidopa extended release (200/50 mg) once daily, safinamide (100 mg per day) and pramipexole extended release (2.1 mg once daily).

STAT-ON™ objective of use: In order to have an objective reporting of his state and affecting motor fluctuations, the use of the STAT-ON™ device was

192 STAT-ON™

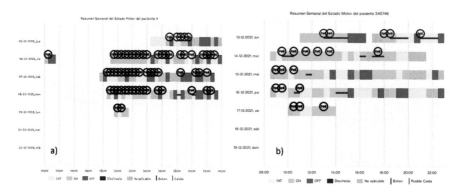

Figure 6.22 STAT-ON™ summary report (A) preapomorphine and (B) postapomorphine.

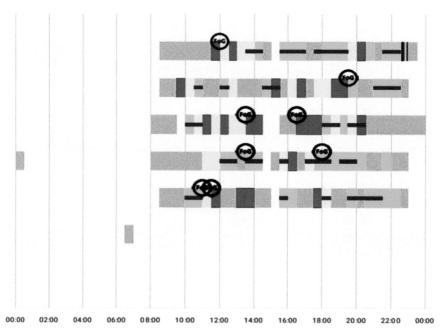

Figure 6.23 STAT-ON™ report summarizes the motor status before starting apomorphine infusion.

decided. Figure 6.23 summarizes the complexity of motor status through the obtained report.

Diagnosis and decision-making: Due to the number of OFF periods and the complexity of motor fluctuations, it was decided to start treatment with subcutaneous apomorphine infusion during the waking day (16 h). The number of hours of inactivity in the report is remarkable, especially in the morning.

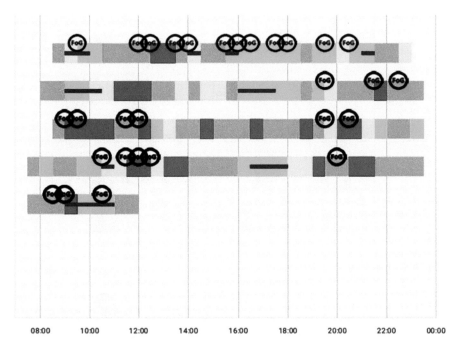

Figure 6.24 Motor status in terms of motor complications after 3 months of subcutaneous apomorphine infusion (with the constant flow).

Figure 6.24 summarizes the motor status in terms of motor complications after 3 months of subcutaneous apomorphine infusion treatment with a flow of 1 mL/h for 16 h. The patient was treated with levodopa/carbidopa immediate release (100/25 mg) six times per day and levodopa/carbidopa extended release (200/50 mg) at bedtime.

Despite apomorphine infusion, motor fluctuations persisted, and nocturnal akinesia was more evident and disabling for the patient. For these reasons, we decided to set the pump with three different flows to improve the motor status of the patient. From 7 am to 12 am the flow was set at 1.2 mL/h, from 12 pm to 11 pm at 1.0 mL/h and from 11 pm to 7 am at 0.4 mL/h.

Figure 6.25 presents the motor status after 6 months of apomorphine infusion and 3 months of variable flows. The patient reported a good response during the night with better overall sleep quality. The motor status improved clearly compared with the baseline in terms of mobility and OFF periods.

Conclusions and take-home messages: In summary, STAT-ON™ is a very useful tool to monitor mobility in advanced Parkinson´s disease patients and can be used as a guide for therapeutic decision-making, including variable flow adjustment of the infusion strategy.

194 STAT-ON™

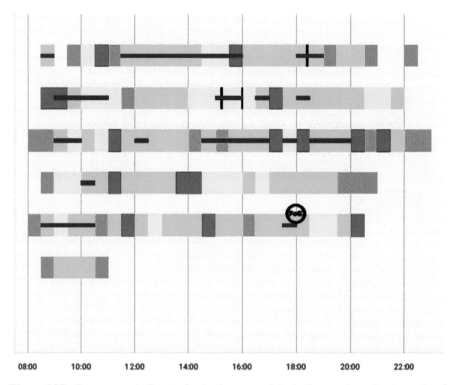

Figure 6.25 Improvement of motor fluctuations, especially in the morning, after 3 months of subcutaneous apomorphine infusion with variable flow.

6.13 Simultaneous Recording of Motor Activity with the STAT-ON™ Device and Subthalamic Nucleus Field Potentials (Percept™) in Parkinson's Disease

Responsible professionals: Dr. José Maria Barrios López and Dr. Lucía Triguero
Hospital Universitario Virgen de las Nieves. Granada.

Personal history: Maternal family history of myasthenia gravis and diabetes mellitus in some relatives and personal history of tonsillectomy.

Parkinson's disease history: A 40-year-old man with advanced juvenile Parkinson's disease (PD) secondary to a homozygous mutation of the PARK2 gene, with a disease course of 34 years. At the age of 6 years, he progressively developed gait impairment due to episodes of dystonia in the left foot. Later, in adolescence, he started with a left-hand tremor. Clinical symptoms improved and remained stable for a few years after starting treatment with levodopa/benserazide and pramipexole.

During the follow-up, a comprehensive workup was performed, including laboratory tests, a normal brain magnetic resonance imaging, and a DaTSCAN which revealed significantly decreased presynaptic dopaminergic transporters in both striatal nuclei. Additionally, a genetic study of dystonia was negative, and finally, a homozygous pathogenic variant of the PARK2 gene was discovered.

During the course of the disease, he began to develop motor complications with simple and complex fluctuations (including delayed ON, wearing-off, and severe OFF state with tremor and asymmetric stiffness predominantly in the left arm and lower limbs, back and leg pain, and inability to walk), peak-dose and possible biphasic dyskinesias.

He was also diagnosed with a psychotic episode related to dopamine agonists. Different treatments were tested, including levodopa/benserazide, pramipexole, ropinirole, rasagiline, and opicapone. Considering the motor complications, treatment was adjusted with levodopa/benserazide (200/50 mg) six times a day, ropinirole retard (2 mg) every other day, and opicapone (50 mg) daily.

Physical examination: Examination in ON state (UPDRS-III 11; Hoenh and Yahr stage II): facial hypomimia ¼, no hypophonia or dysarthria. Mild resting tremor in left arm ¼. No stiffness. Bradykinesia in the left extremities 1-2/4. Choreic/dystonic dyskinesias in the feet. Standing upright is possible without support. Gait with reduced swinging of the left arm. Negative pull test.

Examination in OFF state (UPDRS-III 55; Hoenh and Yahr stage IV): facial hypomimia and hypophonia 2/4. Resting and action tremor predominantly in the left arm and right leg (both ¾). Generalized stiffness, 2/4 axial and ¾ in all four extremities. Global bradykinesia, 2/4 in the right limbs and ¾ in the left limbs. Standing upright is possible without support. Gait with short steps, frequent freezing at start and turn, and choreic/dystonic movements of both feet. Positive pull test (3/4).

Advanced juvenile PD secondary to homozygous mutation of the PARK2 gene with simple and complex motor fluctuations and dyskinesias predominantly in the lower limbs with gait interference was diagnosed. Treatment with bilateral subthalamic nucleus deep brain stimulation (STN-DBS) with the Percept™ neurostimulator (Medtronic) was decided.

STAT-ON™ objective of use: The objective was to describe in clinical practice the simultaneous recording of local field potentials (LFPs) using Percept™ and the motor status using STAT-ON™ in a patient with PD, who underwent bilateral STN-DBS, in order to optimize the treatment.

Once the anatomical location of the electrodes was verified and before starting continuous stimulation, LFPs with Percept™ and motor activity with

STAT-ON™ were recorded for one week. In addition, the patient marked different events: best ON worst OFF; generalized ("dose peak") and leg ("biphasic") dyskinesias; and medication intake. Finally, we analyzed if there was a correlation between the marked events and the recording of both systems.

Diagnosis and decision-making: During the recording period, synchronicity was observed between the events marked by the patient and the results of the STAT-ON™ and Percept™ devices (Figure 6.26). To be emphasized:

- The OFF periods are in agreement with STAT-ON™ OFF state recordings and beta bands.
- Conversely, ON periods are in coincidence with the absence of beta bands, the presence of gamma bands, and non-OFF states reported by STAT-ON™.
- "Dose peak" dyskinesias coincided with dyskinesias identified by STAT-ON™, that is, with gamma bands and without beta bands.
- On the other hand, "biphasic" dyskinesias coincided with a beta band.
- The STAT-ON™ device also detected episodes of FoG, most of them coinciding with "OFF" states and beta bands.

After initiation of bilateral STN-DBS, a decrease in the daily recording of beta bands was observed, coinciding with the disappearance of tremor and stiffness, and substantial improvement in global bradykinesia and gait. On subsequent visits, the stimulation parameters were adjusted, allowing the levodopa/benserazida dose to be reduced and ropinirole to be discontinued, thereby decreasing motor fluctuations and dyskinesias.

Discussion: In our patient undergoing bilateral STN-DBS with the Percept™ system, concordance was observed in the simultaneous recording of motor complications with LFPs and STAT-ON™. In addition, STAT-ON™ was also able to detect FoG and different degrees of motor status. Therefore, this device could be useful in the outpatient monitoring of motor complications in patients with PD treated with DBS in order to optimize therapeutic management.

Conclusions and take-home messages: Outpatient monitoring of motor complications with new technologies is a complementary tool to the anamnesis and clinical evaluation of patients with PD, allowing a more precise assessment of the patient's daily motor status. For example, the recording of LFPs allows more physiological and accurate monitoring, while the

6.13 Simultaneous Recording of Motor Activity

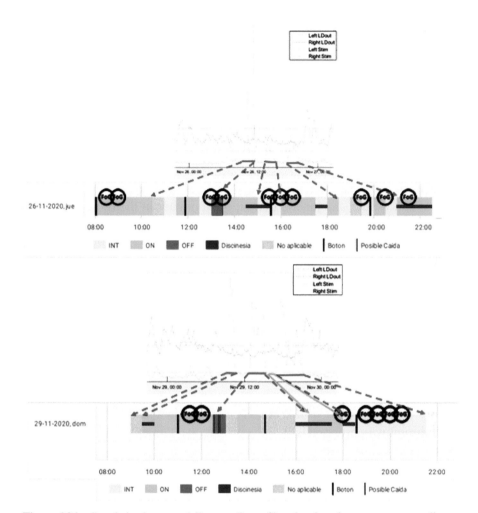

Figure 6.26 Correlation between daily recordings of beta bands and motor status according to Percept™ and STAT-ON™, respectively. Beta bands (blue spikes in the top graph) coincide with OFF (red), "intermediate" (yellow) or "not applicable" (gray = no motion detected) motor status periods detected by STAT-ON™. The ON periods (green) coincide with beta-band free intervals. Most freezing of gait (FoG) episodes detected by STAT-ON™ coincide with non-ON periods.

STAT-ON™ device provides a noninvasive recording and enables the detection of FoG episodes [19, 20].

In our patient with PD who underwent STN-DBS, we confirmed concordance in the indirect recording of motor complications with STAT-ON™. Therefore, the device can be a useful tool in therapeutic optimization in patients who underwent DBS.

6.14 Telemedicine in Parkinson's Disease: The Role of STAT-ON™

Responsible professionals: Dr. Alvaro García-Bustillo and Dr. Esther Cubo
Movement Disorders Unit. Hospital Universitario de Burgos. Burgos.

Personal history: Female, 75-year-old, right-handed, her medical history is significant for arterial hypertension and hypercholesterolemia. From family history, her father had Parkinsonism and dementia at the age of 70 years.

Parkinson's disease history: She consulted at the age of 72 due to a history of rest tremor and kinetic tremor, predominant in the right extremities for one year. She did not have cognitive impairment, predominant dysautonomic symptoms, or early gait impairment. Based on the neurological exam, significant for normal cognitive status, mild bradykinesia, and rigidity, and decreased right arm swing with normal postural responses, she was diagnosed with PD according to the MDS-criteria [21], with a Hoehn and Yahr stage of 2.

She was stable for few years with a good response to levodopa (300 mg/day). In follow-up, 5 years later, her motor status started deteriorating with falls, motor fluctuations, and mild dyskinesias with incomplete response to treatment adjustments. She could not receive dopaminergic agonists due to the presence of mild hallucinations. Based on her risk of falling and unclear history of timing for her OFF periods, she was included in a telemedicine, multidisciplinary program [22] with occupational therapists, nurses, and neurologists to improve balance and motor fluctuations.

Physical examination: The patient was evaluated at a baseline visit and 4 months later. We completed the following assessments: The Movement Disorders Society Unified Parkinson's Disease Rating Scale (MDS-UPDRS) [23], for motor status severity and disability; Freezing of Gait Questionnaire (FoG-Q) [24], and the Mini Balance Evaluation System Test (Mini-BESTest), for balance, postural control, sensory orientation, and dynamic gait assessment; Non-Motor Symptoms Scale (NMSS), for the assessment of the non-motor symptoms; The Parkinson's Disease Questionnaire (PDQ-39), for health-related quality of life. The results of these assessments are shown in Table 6.3.

STAT-ON objective of use: This patient was diagnosed with PD with motor fluctuations, gait impairment, and falls. Given the high risk for hallucinations with an increased dose of dopaminergic drugs, it was decided to include

Table 6.3 Results of the assessments pre and postmultidisciplinary telemedicine program.

	Basal visit	4 months visit	Improvement percentage
MDS-UPDRS (total score)	56	43	23.21
MDS-UPDRS (part III score)	42	31	26.19
FoG-Q	5	4	20.00
Mini-BESTest	21	24	14.29
NMSS	19	15	21.05
PDQ-39	15	13	13.33

* In the MDS-UPDRS, FOG-Q, NMSS and PDQ-39 lower score indicate better status, while in Mini-BESTest lower scores indicate worse status. Improvement percentage was calculated as (final score-baseline score)/baseline score.

her in a multidisciplinary telemedicine program. In this program, the patient received monthly teleconsultations with neurologists and nurses and weekly telerehabilitation sessions with occupational therapists for 4 months.

The objectives were to improve her quality of life by decreasing the risk of falling, increasing physical activity, and improving balance and gait by adjusting her PD medications based on the presence of off periods and disabling dyskinesias.

Additionally, it was decided to monitor her Parkinsonian motor symptoms with STAT-ON™ with the following objectives:

- Adjusting the antiparkinsonian medications based on the timing and duration of the OFF periods.

- Monitoring the frequency of falls and their relationship with the OFF periods.

- Assessing the amount of physical activity as a therapeutic target for physical therapy intervention.

- Evaluating the usability of wearable sensors in patients with advanced PD.

Diagnosis and decision-making: To achieve these goals, she wore the STAT-ON™ device, while she was performing her daily living activities. Baseline and 4-months (after completing the multidisciplinary telemedicine program) assessments provided by STAT-ON™ are shown in Table 6.4.

Discussion: In addition to the PD clinical information provided by the PD rating scales, assessing motor and gait/balance severity, nonmotor symptoms, and quality of life, STAT-ON™ was able to provide additional motor information while the patient was performing her daily living activities.

Table 6.4 Summary of the motor status measured by STAT-ON™ at the baseline visit and after 4 months of the program.

	Baseline visit	4-months visit
Monitored days	8	7
Monitored time (hours)	99.5	85
Number FoG episodes	2	0
Av. FoG episodes/day	0.2 ± 0.4	0 ± 0
Average walking minutes/day	99.7 ± 29.6	108.9 ± 14
Average number steps/day	9959.5 ± 3026.9	10,892 ± 1765.5
Total time in OFF state	39.5 h (39.7%)	28 h (32.9%)
Total time in intermediate state	27 h (27.1%)	19.5 h (22.9%)
Total time in ON state	22.5 h (22.6%)	28.5 h (33.6%)
Total time with dyskinesia	12 h (12.1%)	7 h (8.2%)

With STAT-ON™, we were able to visualize the worst OFF periods, FoG episodes, and the presence of falls and ON periods with dyskinesias. Based on STAT-ON™ reports, we advised her to increase physical activity, and slightly increased the levodopa dose with higher doses in the evening without significantly increasing the hallucinations.

Of note, this patient was satisfied with STAT-ON™ after using it for a relatively long time. The adherence to new technologies and the easiness of using them for patients with advanced age are still controversial. In this case, our patient was living with her husband, who was cognitively intact and eager to use new technologies. We obtained remote information for 4 months, facilitating the PD adjustments based on her motor fluctuations and nonpharmacological interventions, promoting physical activity. However, there is no doubt that the support of her husband and the education provided by the health professionals contributed to overcoming the barriers to using wearables in these populations.

With the clinical information provided by the combination of STAT-ON™ plus standard PD rating scales, we could monitor the treatment response, progression of her disease, and the success of our novel, multidisciplinary telemedicine intervention. In addition, the evaluation of STAT-ON™ reports by the neurologist was not considered high-time consuming.

Conclusions and take-home messages: PD may be considered particularly fitting for distance health/remote assessments with wearable sensors because of the critical importance of the presence, distribution, and characteristics of OFF periods, dyskinesias, and gait impairment. PD patients, especially those with advanced age and living in remote areas, have increased difficulty accessing movement disorder neurologists and other health professionals.

The combination of standard clinical information obtained in in-office consultations, plus remote assessments provided by STAT-ON™, allows better therapeutic management of PD motor symptoms.

6.15 Conclusion

The validation of a medical device by health professionals, during the normal exercise of their activity, is one of the necessary steps to be covered in the acceptation way of a new product introduced into the market. Since 2019, when STAT-ON™ obtained its CE marking as class IIa medical device, the promotion and diffusion activity among the neurologists, hospitals, movement disorders units, and health professionals has been a prominent activity done by the manufacturer.

This chapter has presented a collection of 13 real use cases developed in different Spanish hospitals, using STAT-ON™ as a complementary technology tool that has been used with a diversity of objectives, arriving to determine the usefulness of the device for several reasons (helping the doctor to improve the therapy, identifying candidates to SLT, contributing to a better adjustment of infusion variable dosage, improve the awareness of the patient, complementing or substituting the patient's diary, etc.). The summary of the use cases with the main conclusions is provided in Table 6.5.

References

[1] Kleiner G, Fernandez HH, Chou KL, Fasano A, Duque KR, Hengartner D, et al. 'Non-Motor Fluctuations in Parkinson's Disease: Validation of the Non-Motor Fluctuation Assessment Questionnaire'. Mov Disord. 2021;36(6):1392–1400.

[2] Martínez-Fernández R, Schmitt E, Martinez-Martin P, Krack P. 'The hidden sister of motor fluctuations in Parkinson's disease: A review on nonmotor fluctuations. Mov Disord 2016;31(8):1080–94.

[3] Kim A, Kim HJ, Shin CW, Kim A, Kim Y, Jang M, et al. 'Emergence of non-motor fluctuations with reference to motor fluctuations in Parkinson's disease'. Parkinsonism Relat Disord 2018;54:79–83.

[4] Witjas T, Kaphan E, Azulay JP, Blin O, Ceccaldi M, Pouget J, et al. 'Nonmotor fluctuations in Parkinson's disease: frequent and disabling'. Neurology 2002;59(3):408–13.

[5] Antonini A, Martinez-MartinP, Chaudhuri RK, Marello M, Hauser R, Katzenschlager R, et al. . 'Wearing-Off Scales in Parkinson's Disease:

Table 6.5 Summary of the presented real use cases.

Case	Title	Age	Sex	Center/hospital	Location	Main conclusions and benefits.
2	Early detection of motor fluctuations	65	Male	Centro Médico TEKNON	Barcelona	• Verification of motor fluctuations • Treatment adjustment • Awareness of patient
3	Improving awareness of the first motor fluctuations	59	Male	Complex Hospitalari Moisès Broggi	Sant Joan d'Espí	• Detection of early fluctuations. • Patient's awareness.
4	Complimenting a poor patient's interview about her motor complications	61	Female	Hospital Parc Taulí	Sabadell	• Substitution of patient's diaries. • Detection of OFF states.
5	Indirect detection of probable PD nonmotor fluctuations	67	Female	Complex Hospitalari Moisès Broggi	Sant Joan dEspí	• Help neurologist to identify NMF in association with MF.
6	Deciphering the patient's complaints using STAT-ON™	68	Female	Centro Médico TEKNON	Barcelona	• Precise identification of OFF states and dyskinesia • Patient's education of the knowledge about motor symptoms.
7	Ambulatory monitorization of a patient with advanced PD	70	Female	Hospital Universitario de Toledo	Toledo	• Education of the patient in the relationship between MF and medication intakes. • Decision for an SLT.
8	Improvement of the patient's awareness of the advanced PD stage and the need of second-line treatment	73	Female	Hospital Verge de la Cinta	Tortosa	• Detection of a candidate for an SLT • Objective identification of ON–OFF periods.

9	Identification of candidates to device-aided therapy	58	Female	Complejo Hospitalario Universitario de A Coruña — A Coruña	• Identification of FoG presence • Better adjustment of Duodopa therapy.
10	STAT-ON™ use for LCIG tube adjustment	73	Male	Hospital Universitario Infanta Leonor — Madrid	• Quantification of improvement/worsening of the patient due to LCIG therapy.
11	Monitoring FoG and second-line treatment	69	Male	Complexo Hospitalario Universitario de Pontevedra — Pontevedra	• Complementary information to the Hauser diary • Monitoring a SLT results.
12	Improving motor fluctuations with variable flow of apomorphine subcutaneous infusion. The role of STAT-ON™	73	Male	Campus Universitario Vall d'Hebron — Barcelona	• Therapeutic decision-making • Contribution to the variable flow adjustment of a subcutaneous infusion strategy.
13	Simultaneous recording of motor activity with the STAT-ON™ device and subthalamic nucleus field potentials (PERCEPT™) in PD	40	Male	Hospital Universitario de Burgos — Burgos	• Confirmation of the relationship between MF (recorded with STAT-ON™) and the LPF (recorded with Percept™) in a DBS-implanted patient.
14	Telemedicine in PD. The role of STAT-ON™	75	Female	Hospital Universitario de Burgos — Burgos	• Contribution to the improvement of the therapeutic management of advanced age PD patients, living in rural areas.

Critique and Recommendations'. Mov Disord 2011 Oct;26(12):2169–75. doi: 10.1002/mds.23875. Epub 2011 Jul 20.

[6] Goetz CG, Stebbins GT, Chmura TA, Fahn S, Poewe W and Tanner C. 'Teaching program for the Movement Disorder Society-Sponsored Revision of the Unified Parkinson's Disease Rating Scale: (MDS-UPDRS)'. Movement Disorders 2010; Jul 15;25(9):1190–4. doi: 10.1002/mds.23096.

[7] Hauser RA, Friedlander J, Zesiewicz TA, Adler CH, Seeberger LC, O'Brien CFO, et al. 'A home diary to assess functional status in patients with Parkinson's disease with motor fluctuations and dyskinesia'. Clin Neuropharmacol 2000; 23(2):75–81. doi: 10.1097/00002826-200003000-00003.

[8] Espay A, Bonato P, Nahab FB, Maetzler W, Dean JM, Klucken J, et al. 'Technology in Parkinson's disease: challenges and opportunities'. Mov Disord 2016 Sep;31(9):1272–82. doi: 10.1002/mds.26642. Epub 2016 Apr 29.

[9] Pérez-López C, Samà A, Rodríguez-Martín D, Moreno-Aróstegui JM, Cabestany J, Bayés À, et al. 'Dopaminergic-induced dyskinesia assessment based on a single belt-worn accerelometer'. Artif Intell Med 2016 Feb;67:47–56. doi: 10.1016/j.artmed.2016.01.001. Epub 2016 Jan 14.

[10] Rastgardani T, Armstrong MJ, Gagliardi AR, Marras C. 'Understanding, Impact, and Communication of "Off" Periods in Parkinson's Disease: A Scoping Review'. Mov Disord Clin Pract. 2018 Oct 9;5(5):461–470. doi: 10.1002/mdc3.12672. PMID: 30515435; PMCID: PMC6207105.

[11] Montanaro E, Artusi CA, Zibetti M, Lopiano L. 'Complex therapies for advanced Parkinson's disease: what is the role of doctor-patient communication?' Neurol Sci. 2019 Nov;40(11):2357–2364. doi: 10.1007/s10072-019-03982-5. Epub 2019 Jun 28. PMID: 31254180.

[12] Antonini A, Stoessl AJ, Kleinman LS, et al. 'Developing consensus among movement disorder specialists on clinical indicators for identification and management of advanced Parkinson's disease: a multi-country Delphi-panel approach'. Curr Med Res Opin 2018;34:2063–73.

[13] Martinez-Martin P, Kulisevsky J, Mir P, Tolosa E, García-Delgado P, Luquin MR. 'Validation of a simple screening tool for early diagnosis of advanced Parkinson's disease in daily practice: the CDEPA questionnaire'. NPJ Parkinsons Dis. 2018 Jul 2;4:20.

[14] Santos García D, López Ariztegui N, Cubo E, et al. 'Clinical utility of a personalized and long-term monitoring device for Parkinson's disease in a real clinical practice setting: An expert opinion survey on STAT-ON™'. Neurologia (Engl Ed) 2020:S0213–4853(20)30339-X.

[15] Bougea A, Palkopoulou M, Pantinaki S, Antonoglou A, Efthymiopoulou F. 'Validation of a real- time monitoring system to detect motor symptoms in patients with Parkinson's disease treated with Levodopa Carbidopa Intestinal Gel [abstract]'. Mov Disord (2021). Available online at: https://www.mdsabstracts.org/abstract/validation-of-a-real-time-monitoring-system-to-detect-motor-symptoms-in-patients-with-parkinsons-disease-treated-with-levodopa-carbidopa-intestinal-gel/ (accessed May 12, 2022).

[16] Barrios-López JM, Ruiz Fernandez E, Triguero Cueva L, et al. 'Registro simultáneo de la actividad motora con sensores inerciales (STAT-ON™) y de potenciales de campo de núcleo subtalámico (PerceptTM) en la enfermedad de Parkinson'. XLIII Reunión Anual Sociedad Andaluza Neurología Sevilla (2021).

[17] Olanow CW, Kieburtz K, Odin P, Espay AJ, et al. 'Continuous intra-jejunal infusion of levodopa-carbidopa intestinal gel for patients with advanced Parkinson's disease: a randomised, controlled, double-blind, double-dummy study'. Lancet Neurol. 2014 Feb;13(2):141–9. doi: 10.1016/S1474-4422(13)70293-X.

[18] Hubert H Fernandez, James T Boyd, Victor S C Fung, et al. Long-term safety and efficacy of levodopa-carbidopa intestinal gel in advanced Parkinson's disease. Mov Disord. 2018 Jul;33(6):928–936. doi: 10.1002/mds.27338.

[19] Neumann W-J, Staub F, Horn A, Schanda J, Mueller J, Schneider G-H, et al. 'Deep brain recordings using an implanted pulse generator in Parkinson's disease'. Neuromodulation. 2016;19(1):20.

[20] Neumann W-J, Staub-Bartelt F, Horn A, Schanda J, Schneider G-H, Brown P, et al. 'Long term correlation of subthalamic beta band activity with motor impairment in patients with Parkinson's disease'. Clin Neurophysiol. 2017;128(11):2286.

[21] Postuma RB, Berg D, Stern M, Poewe W, Olanow CW, Oertel W, et al. 'MDS clinical diagnostic criteria for Parkinson's disease'. Mov Disord [Internet]. 2015 [cited 2022 Oct 28]; 30(12): 1591–1601. Available from: https://doi.org/10.1002/mds.26424

[22] Cubo E, Garcia-Bustillo A, Arnaiz-Gonzalez A, Ramirez-Sanz JM, Garrido-Labrador JL, Valiñas F, et al. 'Adopting a multidisciplinary telemedicine intervention for fall prevention in Parkinson's disease. Protocol for a longitudinal, randomized clinical trial'. PLoS One [Internet]. 2021 [cited 2022 Oct 28]; 16(12). Available from: https://doi.org/10.1371/journal.pone.0260889

[23] Goetz CG, Tilley BC, Shaftman SR, Stebbins GT, Fahn S, Martinez-Martin P, et al. 'Movement Disorder Society-sponsored revision of the

Unified Parkinson's Disease Rating Scale (MDS-UPDRS): scale presentation and clinimetric testing results: MDS-UPDRS: Clinimetric Assessment'. Mov Disord [Internet]. 2008 [cited 2022 Oct 28]; 23(15): 2129–2170. Available from: https://doi.org/10.1002/mds.22340

[24] Giladi N, Tal J, Azulay T, Rascol O, Brooks DJ, Melamed E, et al. 'Validation of the freezing of gait questionnaire in patients with Parkinson's disease: Validation of FOG Questionnaire'. Mov Disord [Internet]. 2009 [cited 2022 Oct 28]; 24(5): 655–661. Available from: https://doi.org/10.1002/mds.21745

7

New Open Scenarios for STAT-ON™: The Medical Perspective

Núria Caballol[1,2] and Diego Santos-Garcia[3,4]

[1]Hospital de San Joan Despí, Departament de Neurologia, Complex Hospitalari Moisès Broggi, Sant Joan Despí, Spain
[2]Unitat de Parkinson i Transtorns de Moviment, Centro Médico TEKNON, Barcelona, Spain
[3]CHUAC – Complejo Hospitalario Universitario de A Coruña, A Coruña, Spain
[4]Departamento de Neurología, Hospital San Rafael, A Coruña, Spain

Abstract

The chapter covers a transversal vision of the possible new scenarios where STAT-ON™ can positively contribute to helping professionals in the development of the clinical activity, and generate new possibilities in the treatments and patients' management.

7.1 Introduction

This chapter aims to provide a cross-sectional view of the possible scenarios for the use of the Holter STAT-ON™, from the perspective of medical practice, in the treatment and follow-up of patients with Parkinson's disease (PD). Today, there exists a practical unanimous agreement on the advantages and contributions that the proper use of technology implies in various aspects of our lives, such as health care. In the case of Parkinson's disease, and for reasons inherent to the disease itself, this possibility has been a little further from being able to become a reality, due to the nonavailability of the most suitable technology.

The STAT-ON™ solution opens up a good number of opportunities to make this contribution to improving the care and supervision of PD patients effectively, always as a technological complement that provides objective and reliable information on the patient's motor status and its evolution, allowing the doctor to have a very correct vision of the patient's condition, in normal living conditions, thus going beyond the information that the doctor can observe in his office, at the time of the visit, or that can be provided by the patient himself, which on some occasions may be biased, qualitative, or imprecise.

In Chapter 6 several real cases were presented, corresponding to patients affected by PD and treated in different Spanish hospitals. In all cases, a presentation has been made of the contribution that the use of information obtained using STAT-ON™ has meant for the case. On many occasions, these benefits have been translated into better monitoring of the evolution of the disease, in establishing the appropriate criteria for a change in treatment or in improving the patient's own ability to become aware of the disease itself, allowing to establish a much more fruitful relationship with the neurologist.

The extrapolation of a series of cross-sectional conclusions has represented the possibility to establish the content that follows in the hereafter sections, and which leads to establishing the appropriate framework to glimpse a series of uses and future applicability of the STAT-ON™ device.

STAT-ON™ is conceived and marketed as a medical device for the detection and measurement of motor symptoms associated with PD. Therefore, it appears within the product specifications, it is capable of correctly detecting the appearance of dyskinesias, OFF states, and the presence of Freezing of Gait (FoG). However, as discussed below, the use that the professional can make of this information may be novel (for example, due to the existing correlation between the OFF states of the patient and the possible associated nonmotor-motor fluctuations (NMFs)).

Even though STAT-ON™ was conceived, tested, and proven from a database that included patients affected by Parkinson's who presented motor fluctuations (MFs) and were in intermediate stages of disease progression, it has been verified by the neurologists who have been using it that it can be useful in detecting the first MFs (which therefore affect inexperienced patients and who may have significant difficulty in describing them to their neurologists).

It has also been seen the importance that this technology can have in the correct identification of patients who are candidates to be users of technologically-assisted treatments (deep brain stimulation (DBS), infusion pumps, etc.), and their subsequent correct follow-up and necessary adjustment. Finally, STAT-ON™ can be a definitive aid for the more effective execution of clinical trials that require the participation of patients by filling in Hauser's diaries or personal symptom diaries. The sensor and the generated report are good candidates to be an "electronic and automatic diary of the associated motor symptoms."

7.2 Detection of the First PD Motor Fluctuations

One of the main purposes of the use of the STAT-ON™ sensor is to help clinicians in the detection of the motor fluctuations (MFs). It is well-known, that traditional methods, such as a detailed clinical interview, validated clinical scales, or patient diaries can be useful [1–3]. Nonetheless, not all patients are always aware of their OFF time [4]. Some of the clinical cases described in Chapter 6, illustrate the real difficulties of PD neurologists when the clinicians need to go deeply into the MF details and the patients lack awareness of them.

The detection of the first MF can be a challenge, especially in the first years of the disease. The recommended wearing-off (WO) scales such as the 19-item *WO* or *Quick* questionnaires can increase the detection of WO in the setting of the daily clinical practice. However, it is not always possible for PD neurologists to use them in the daily-clinical practice scenario, mainly for the lack of time for each patient. Besides, the first appearance of MF (morning akinesia and WO) can be very subtle at the very beginning. The transition from a good motor state to a worst one can be gradual and ambiguous [1, 5]. While some patients can quantify and identify WO symptoms quite well, for others it is extremely difficult. Some studies that analyze the perception gaps between patients and physicians in terms of the detection of motor complications, show a lack of awareness of WO among PD patients [4].

Otherwise, several communication barriers can exist when explaining WO, such as patient's cognitive impairment, reluctance to discuss the symptoms, or caregiver absence [6]. To overcome all these challenges and barriers, wearable-sensor-based technology can help physicians to detect WO [7, 8]. Since the introduction of the STAT-ON™ in June 2019 in the setting of Spanish centers, PD neurologists are using this wearable to quantify the patient's OFF time [9]. However, the STAT-ON™ has been more widely used in advanced PD patients so far, being less explored the first phase of PD where the first MF emerges [9].

Even considering that the algorithms implemented in the sensor for the determination of the ON and OFF states and therefore the MFs were built with patients with advanced PD, and fully aware of their ON/OFF state (i.e., the learning database does not include patients who are not able to know and identify their motor status) [10], our preliminary results with the applicability of the STAT-ON™ to detect the first MF are very encouraging [5, 11–13].

In a retrospective analysis of 35 PD patients with a mean disease duration of 4.07 ± 1.0 years at the time of wearing the sensor, WO or morning-akinesia (MA) was suspected to occur by the neurologist but they were not well recognized by the patient or caregiver in 40% of the sample. Moreover, in 33.3% of the PD patients WO and MA were not suspected before wearing the sensor. After wearing the STAT-ON™ sensor, patient's and caregiver's self-awareness increased because in all the patients in whom WO/MA was not well recognized, the STAT-ON™ report showed the presence of them in all the cases. Among the patients who denied having MF/MA, the sensor detected them in 9 out of 10 patients. There were 6 cases in whom the STAT-ON™ report helped the patient/caregiver to understand that MF was occurring and recognized them. Still, three patients denied again having WO after wearing the sensor, but the STAT-ON™ showed it. In this line, another retrospective analysis focusing only on the MA of 28 PD patients, showed significant differences of the morning gait fluidity between patients without clinically suspected MA and those with MA clinically present and well explained and quantified for the patients [13].

All these preliminary results suggest that STAT-ON™ sensor is a promising and helpful tool for the neurologists who want to confirm the occurrence of the first MF. However, a critical issue that will need further study and clarification is the explanation for a disagreement between the symptoms and the sensor (i.e., when the sensor is detecting OFF or MA and the patient denies them). Besides, the well-known lack of awareness of the symptoms among patients or the simple fact that they minimize the symptoms, our hypothesis is that the sensor has a high sensitivity to detect a slowness of the gait fluidity before and after the levodopa intake. Hence, the sensor may be detecting this OFF transition before the patient is aware of it [5]. More studies in larger PD samples, addressing all these issues should be necessary.

7.3 Identification of Freezing of Gait and Falls

FoG is a frequent and disabling symptom in PD and a major risk factor for falls [14]. It is defined as sudden and usually brief episodes of inability to produce effective forward stepping that clinically occur during gait initiation or turning [15]. Detection of FoG is extremely important for PD neurologists for several reasons:

- Firstly, for classifying the type of parkinsonism and rule out and atypical parkinsonism or to identify a postural instability/gait difficulty

7.3 Identification of Freezing of Gait and Falls   211

PD subtype, which is associated with a faster cognitive and motor decline [15].

- Secondly, to explore if the FoG is occurring in the ON or the OFF state, especially when selecting PD patients for DBS.

- Thirdly, for the implementation, as soon as possible, of a more adequate therapy, either pharmacological or nonpharmacological, and consequently try to prevent falls.

- Fourthly, to measure FoG reduction after pharmacological/nonpharmacological therapies. As seen in several cases of the previous chapter, it is well illustrated that FoG improves after initiating dopaminergic therapies.

- Finally, to detect longitudinally the appearance and progression of this disabling symptom that complicates the course of PD.

Besides the clinical information provided by the patient and the specific FoG scales, the sensor can serve as a complementary tool to detect the presence of FoG. In line to the detection of the first MF, the detection of the first FoG episode can also be a challenge. While in some patients FoG can be identified at the clinical examination or during the clinical interview, in some other instances the lack of time of the clinician for a proper identification of all the PD symptoms, can produce an underdiagnose of FoG. Moreover, another factor that accounts for an underdiagnosed FoG is that the patient's examination only reflects the motor state at a precise moment, without reflecting all the PD symptoms along a day [7].

For these reasons, wearable sensors such as STAT-ON™ are of great help. Our first experience **using the STAT-ON™ sensor for detecting FoG was that the sensitivity of the sensor was again extremely high**.

In our experience, in some cases it could be recommended to have "a daily-activities diary" while wearing the sensor (as it has been seen in the cases of Chapter 6). The use of this diary makes possible a more precise interpretation of the results when there is a possible disagreement (i.e., the STAT-ON™ reports a FoG but the patient denies its presence). We know that daily activities such as sweeping or "stop walking suddenly" can be detected as a FoG by the sensor, generating a "FoG false positive" that can be discarded or interpreted if such "daily-activities diary" is available from the patient. Still, a recently published pilot study with the STAT-ON™ in an unsupervised scenario showed that a 76.9% of agreement between the clinical interview and the FoG was detected by the sensor with a kappa coefficient of 0.481 [5]. However, in this study FoG specific clinical scales were not applied.

In summary, using the STAT-ON™ sensor for all the purposes commented is worthwhile and several projects addressing these issues are ongoing.

7.4 Detection of Dyskinesias

Although recent and controversial views on the management of PD have suggested an overall decline of dyskinesia rates, the detection of dyskinesia is one of the main objectives when evaluating the PD patient [16]. Several reasons such as a more conservative use of levodopa regimes, the earlier introduction of DBS and other device-aided therapies, can account for the decline of these dyskinesia rates [16]. Nevertheless, and despite the negative results of a set of anti-dyskinetic agents, the clinical trials in the field of anti-dyskinetic agents are ongoing and STAT-ON™ could be of help as explained in this section.

STAT-ON™ has some limitations and cannot detect all the types of dyskinesia, for all the time because the sensor only detects the dyskinesia when the patient is not walking. Besides, the sensitivity and specificity are of 95% and 93% for strong or mild trunk dyskinesia while for mild upper limb dyskinesia the sensitivity is lower (39%) [17].

Despite that, our first experience using the sensor in clinical practice is that the STAT-ON™ can help the physician to improve the patient's awareness of dyskinesia. In several cases of the previous Chapter 6, it is shown that some patients mix up tremor and dyskinesia symptoms. Although the STAT-ON™ sensor cannot be used to detect tremor, it can be very helpful to detect dyskinesia when the patient thinks that "a dyskinesia confused as a tremor" is emerging. The procedure, then, can be to ask the patient to press the button of the STAT-ON™ when this situation is appearing.

In line with the previous aforementioned "false positives" possibilities with FoG (see section 7.3), in the case of dyskinesias, dancing, and housework are activities that can be easily confused with dyskinesia. For this reason, when discussing the report with the patient, it is important to review the patient's "daily-diary activities" to detect disagreements between the sensor and the patient.

Despite these challenges, detecting the first dyskinesias is a critical point during PD (similar to the detection of initial MF) because it indicates that the patient is entering a more complex phase of the disease and treatment adjustments are mandatory from this moment. Besides the use of the STAT-ON™ for measuring the effect of the treatment interventions, the sensor can also help the clinician in the detection of the dyskinesias. However, further studies should be addressed to confirm the ability of the sensor in detecting the different types of PD dyskinesia.

7.5 Detecting Non-motor Fluctuations

In principle, many of the new wearable devices for monitoring the PD patients are focused on the detection of the related motor symptoms and, in general, they are considered as useless tools for detecting NMFs.

Despite the fact that STAT-ON™ is not designed to detect nonmotor symptoms (NMS), if the possibility of pressing the button to indicate a certain event is used correctly, it is possible to associate the presence of a certain period or motor state (e.g., an OFF state) with the onset of a state "in which the patient does not feel well" and which is normally associated with a nonmotor symptom condition. In this way, the patient could be instructed to tight the button when NMS appear, and it could be seen in the generated report if those moments have a correlation with detected OFF episodes. Then, STAT-ON™ could be used to identify NMF in an indirect way.

Our very preliminary experience suggests that when the patient complains of a NMF, the STAT-ON™ shows an OFF time detected [12]. This issue has also been illustrated in the previous Chapter 6 and future studies with STAT-ON™, analyzing the characteristics of the patient's OFF time, should be accompanied with the recommended clinical scales that complement the nonmotor part of the OFF state.

7.6 Selection of a Patient for a Device-aided Therapy and Monitor Response

Patients with Parkinson´s disease (PD) develop clinical fluctuations and their identification is very important because these patients have a worse status in terms of motor symptoms, non-motor symptoms (NMS), quality of life (QoL), and autonomy for activities of daily living (ADL) [18]. For example, in the DEEP study [19], wearing-off was detected by the neurologists in more than 20% of the patients in the subgroup with fewer than 2.5 years of disease duration, while with the WOQ-19, 41.8% of patients were detected. Motor symptoms and NMS can be present during the OFF episodes [20] and different instruments could be useful for detecting clinical fluctuations in PD: (1) asking directly to the patient; (2) scales or questionnaires; (3) ON–OFF diaries; (4) wearable tools; (5) website applications; (6) video recording (at home or at the consult).

In this context, effective management of PD is critical at all stages of disease, requiring individual customization of therapy including optimization of oral regimens and consideration for nonoral treatments such as advanced device-aided therapies (i.e., DBS, levodopa infusion, and/or apomorphine infusion) [21]. A lack of consensus around the definition of advanced disease leads to delays in the identification of advanced PD patients, and the correct managing disease progression and its timely treatment [22]. Some tools have been proposed to identify patients inadequately controlled on oral medications such as de 5-2-1 criteria [23–25], the CDEPA questionnaire [26], and the MANAGE-PD [27]. Specifically, the MANAGE-PD provides information about if a patient could be a candidate for a device-aided therapy and despite some limitations its clinical use may complement clinician treatment decision-making and facilitate timely identification and management of PD symptoms [27, 28].

STAT-ON™ could be a tool to help the neurologist when deciding if the patient is a candidate for device-aided therapy as well [29]. Many factors are relevant for deciding if a patient is a candidate for a device-aided therapy. One of the most significant is the total time the patient is in OFF state during the waking day. OFF episodes can be detected with the STAT-ON™ with high sensitivity. The ON/OFF algorithm was also validated against the Hauser diary showing a greater compliance (37% records more were achieved by the sensor) and a high accuracy (positive predictive value 0.92; negative predictive value 0.94) [30].

Recently, a subanalysis of the MoMoPA-EC clinical trial showed a moderate concordance between the STAT-ON™ and the patient diary, but the correlation between the different UPDRS indices was better with the STAT-ON™ than with the Hauser diary [31].

In 2020, Santos-García et al. published the opinion of 27 clinical experts on PD about the STAT-ON™ after having tested the device in clinical practice [32]. A total of 119 evaluations were performed and the STAT-ON™ was considered better than diaries and a useful tool to detect advanced PD by 70.3% and 81.5% of the neurologists, respectively. Moreover, other important signs that can be appearing or increased during the OFF episodes can be detected with the STAT-ON™ such as bradykinesia, freezing of gait (FoG), and falls [33]. Time with dyskinesia is detected with the STAT-ON™ as well.

Some NMS such as pain, fatigue, bad mood, or anxiety can appear during the OFF episodes and improve with a device-aided therapy, so their identification is of great importance [34–36]. These NMS can be indirectly detected with the STAT-ON™ by asking the patient to press the button of the sensor (see above section). All the information collected with the STAT-ON™,

7.6 Selection of a Patient for a Device-aided Therapy and Monitor Response

together with other factors that are considered when deciding if the patient is a candidate for device-aided therapy (i.e., age, family support, comorbidity, cognitive function, etc.), should finally be taken into account.

In summary, the STAT-ON™ could help the neurologist to know about 3 key factors for deciding if a patient is a candidate for a device-aided therapy:

- Time in the OFF state during the waking day.
- Symptoms during the OFF episodes (correlation between records and clinical assessment).
- Severity of OFF episodes (correlation between records and clinical assessment). Specifically, it must be considered the obtained data about OFF episodes, dyskinesia, FoG, falls, and gait problems.

If the patient is finally treated with a device-aided therapy, the STAT-ON™ could be useful for monitoring the response with the new therapy, both in the short- and the long-term. The comparison of records (before vs. after starting with the device-aided therapy) will allow the neurologist to observe the reduction in OFF time and the changes in other variables, as well as long-term monitoring of the patient's condition and the identification of possible therapy adjustments.

The device has been also validated with advanced-stage PD patients with levodopa-carbidopa intestinal gel (LCIG). Bougea et al. demonstrated the better detection of ON/OFF motor fluctuations, dyskinesia, and falls against patients' diaries in 51 patients with advanced PD [37]. All the sensitivities and specificities were higher with the sensor rather than with the diary, concluding that STAT-ON™ can be a promising tool for monitoring patients with advanced disease.

In another study, the STAT-ON™ was used to monitor a patient with LCIG therapy whose motor symptoms were worsening after 4 months of using LCIG therapy. STAT-ON™ was used to check his state and it was detected a bad adjustment of the LCIG tube. After the correct adjustment, the STAT-ON™ was used again to check the improvement on motor states [38].

STAT-ON™ was also used in patients who were administered with PERCEPT™, a deep brain stimulator that also registers the signal perceived from the subthalamic nucleus field, remarkably aligning their signals in the appearance of OFF states, ON states, dyskinesia, and FoG episodes. This case study suggests that STAT-ON™ can be a useful tool for the optimization of this kind of therapy [39].

Another pilot analysis conducted in 11 PD patients, 4 of them with deep brain stimulation, suggested that STAT-ON™ could be useful to have an objective measurement of the motor status of patients in advanced stages of the disease, with difficulty in controlling motor symptoms, inconsistencies in their daily reports, suspicion of inappropriate taking of medication, and in those who were enrolled to a treatment of greater complexity such as surgery [40]. In patients under apomorphine infusion the STAT-ON™ can be a useful tool [41] and in the future could be used in patients receiving new drugs such as subcutaneous levodopa infusion or drugs for rescuing the patient from the OFF state with the aim of monitoring the response. In line with this, other projects using the STAT-ON™ in advanced PD patients, such as the GATEKEEPER project, are ongoing [42].

7.7 Monitor the Response to a Treatment

As it has been already explained and commented in above text, the STAT-ON™ can be used for monitoring the response of a device-aided therapy or other drugs prescribed with the aim to reduce the OFF time in a patient with PD. Clinical fluctuations are very frequent [43] and many options are available for trying to optimize the status of the patient [44]:

- to adjust levodopa,
- to add a dopamine agonist,
- to add a catechol-O-methyltransferase inhibitor, and
- to add a Mmonoamine oxidase-B inhibitor.

In practice, many patients receive all these drugs added to levodopa and it could be helpful and of great interest to monitor with the STAT-ON™, the motor status of the patient before and after starting with the drug selected.

A correlation between the perception of the patient about the effect of the drug over the symptoms and the change in the record collected with the STAT-ON™ would be important information. Even regular monitoring could be used for trying to be more sensible to detect changes in the status of the patient over the time with the aim of adjust the medication early. In all these cases the focus should be the OFF state and symptoms related to the OFF episodes.

A very interesting alternative could be to use STAT-ON™ to monitor dyskinesia in PD patients. The effect of amantadine or other adjustments and/or therapies [44] conducted with the aim of improving dyskinesia could be monitored, again with a comparison between the record before and after starting with the drug. Specific disruptive complications for the patient such as FoG or falls could be also monitored, before and after an intervention with a drug or a dispositive (visual clues, etc.) or other type (physiotherapy, etc.).

Even in PD patients without motor fluctuations the STAT-ON™ could be useful for monitoring the effect of exercise programs over aspects related to gait and daily physical activity. Finally, and very importantly, the use of new wearable sensors such as the STAT-ON™ could be especially useful in patients with mild cognitive impairment or dementia [45], since the data collected with the Hauser diary may be unreliable in these patients.

7.8 Use in Clinical Trials

Currently, there is an ongoing trial with the STAT-ON™ entitled MoMoPA-EC [46]. The objective of this trial is to show whether using the Parkinson Holter (STAT-ON™) is better than the clinical interview used in traditional clinical practice (primary objective), and whether it is not inferior to the ON-OFF diary recorded by the patients at home (exploratory objective). This is a multicenter (43 centers from Spain) randomized with parallel assignment and triple masking (participant, investigator, and outcomes assessor). The primary outcome is the change from baseline to the last visit in daily hours in the OFF state.

Regarding all previous comments about the STAT-ON™ and Hauser diaries, it could be of great interest the progressive introduction of the STAT-ON™ as a tool for measuring the change in the OFF time in those clinical trials conducted in PD patients with MFs. The change in the OFF time from the baseline visit to the final visit measured with the STAT-ON™ could be used as the primary endpoint in trials with a drug for reducing the daily OFF time. Its use would be easier for the patients and the possibility of recording the data even about all the days without a fatigue effect compared to the diaries would be a positive point. Depending on the trial and the endpoint, other variables could be monitored: time with dyskinesia, presence of FoG, falls, etc. Moreover, and regarding advanced PD and symptomatic interventions, the STAT-ON™ could be used for monitoring the effect of device-aided therapies in the context of clinical trials.

Another possibility to explore could be the use of the STAT-ON™ in clinical trials with molecules with a potential modified disease effect. In some trials (drug vs. placebo) the time to motor worsening or the time to starting with levodopa are included as secondary variables and the STAT-ON™ could provide a more objective information about the progression of motor symptoms. Even in open label trials with some therapies (grow factors, cell therapy, etc.), the development of some complications such as OFF episodes,

FoG, falls, or dyskinesia could be defined as endpoints in the very long-term follow-up. In general, the use of new technology is the rule in trials with a disease-modifying molecule with the aim to know more precisely the changes that occur in the disease and be able to compare between placebo and the drug [47].

Finally, the STAT-ON™ could be used as a helpful tool not only in double-blinded clinical trials but also in phase IV studies providing information about clinical real-world evidence. Data about real-world evidence is of great importance after starting with the commercialization of a novel molecule and the use of wearable sensor platforms, smartwatches equipped with accelerometers and other wearable devices could be used for getting very important information [45, 48–53].

7.9 Use as a Marker of Disease Progression

As it has been previously commented, new devices can be used to monitor the progression of the disease. The development of MFs can be considered a turning point in the story of a patient with PD because some therapies in fact are indicated only in patients with levodopa and fluctuations [44]. The STAT-ON™ could be used for trying to detect early predictable fluctuations (morning akinesia; wearing-off), and also in the long-term, to monitor an increase of the daily OFF time or the development of unpredictable motor fluctuations such as no-ON or partial-ON during the afternoon/evening.

Keeping in mind the concept of motor fluctuations development as a turning point, an interesting idea could be to compare the record collected with the STAT-ON™ in early PD patients (i.e., <5 years of disease duration from symptoms onset) in patients with positive vs. negative 5-2-1 criteria [23, 24]. Moreover, the detection and monitor progression of other complications such as FoG, falls, or dyskinesia could be conducted with the STAT-ON™ in the medium- and long-term.

7.10 Research and Future Scenarios with STAT-ON™

The focus of remote technologies is now slowly shifting toward the broad, but more "hidden," spectrum of NMS [54]. To apply technologies in prospective cohorts [55] with the aim of obtaining very valuable data seems to be an interesting approach. Recent clinical research provides growing evidence that various NMS such as neuropsychiatric, autonomic, and sensory symptoms (particularly pain) also show fluctuations in patients with motor fluctuations (called NMF) [20]. This aspect cannot be directly assessed specifically with the STAT-ON™, but it is also known that NMS burden is greater in PD patients with motor fluctuations [18] and the relationship between NMS and motor fluctuations can be explored with the STAT-ON™ combined with data collected applying validated scales (e.g., Non-Motor Symptoms Scales, Non-Motor Fluctuation Assessment Questionnaire, etc.) [56].

Integration of the telemedicine in the management of PD could be useful to remotely monitor the PD motor complications, facilitate the access to care, complement, or replace the in-office consultations specially when these are not possible for geographical reasons, improving the detection of patients who are candidates to a device-aided therapy and facilitate the monitoring of device-aided therapies [57–59]. Moreover, the STAT-ON™ can be used as a part of a multidisciplinary telemedicine intervention with the aim of reducing the risk of some complications such as falls [60]. Even the information of the STAT-ON™ can be complementary to the other devices [61] and all together useful for making decisions about the treatment of the patient with PD. In the future, it would be of great interest to apply the use of the STAT-ON™ and other devices in longitudinal follow-up cohorts [55].

7.11 Conclusion

It can be stated that the role of STAT-ON™ complementary technology is clear, providing objective information on the motor state of PD patients, which the neurologists can use to complement their own observations, helping them to make decisions, in many cases, much more substantiated. This enables a more accurate prescription, directly impacting in the QoL of the patient. The use of STAT-ON™ in clinical practice for evaluating better PD patients, for selecting better and earlier the candidate patients to specific therapies, the use in clinical trials seems to be scenarios where STAT-ON™ fills a gap which seems to be beneficial for the clinician and for the patient. Additionally, the continued use and the experience acquired by a significant number of neurologists give rise to being able to define future and new fields of application.

References

[1] Antonini A, Martinez-Martin P, Chaudhuri RK, Merello M, Hauser R, Katzenschlager R, et al. 'Wearing-off scales in Parkinson's disease: Critique and recommendations'. Mov Disord. 2011;26: 2169–2175. doi:10.1002/mds.23875

[2] Stacy M. 'The wearing-off phenomenon and the use of questionnaires to facilitate its recognition in Parkinson's disease'. J Neural Transm. 2010;117: 837–846. doi:10.1007/s00702-010-0424-5

[3] Hauser RA, Friedlander J, Zesiewicz TA, Adler CH, Seeberger LC, O'Brien CF, et al. 'A Home Diary to Assess Functional Status in Patients with Parkinson's Disease with Motor Fluctuations and Dyskinesia'. Clin Neuropharmacol. 2000;23: 75–81. doi:10.1097/00002826-200003000-00003

[4] Ogura H, Nakagawa R, Ishido M, Yoshinaga Y, Watanabe J, Kurihara K, et al. 'Evaluation of Motor Complications in Parkinson's Disease: Understanding the Perception Gap between Patients and Physicians'. Aasly J, editor. Parkinsons Dis. 2021;2021: 1–8. doi:10.1155/2021/1599477

[5] Caballol N, Bayés À, Prats A, Martín-Baranera M, Quispe P. 'Feasibility of a wearable inertial sensor to assess motor complications and treatment in Parkinson's disease'. Suzuki K, editor. PLoS One. 2023;18: e0279910. doi:10.1371/journal.pone.0279910

[6] Armstrong MJ, Rastgardani T, Gagliardi AR, Marras C. 'Barriers and facilitators of communication about off periods in Parkinson's disease: Qualitative analysis of patient, carepartner, and physician Interviews'. Lawson RA, editor. PLoS One. 2019;14: e0215384. doi:10.1371/journal.pone.0215384

[7] Monje MHG, Foffani G, Obeso J, Sánchez-Ferro Á. 'New Sensor and Wearable Technologies to Aid in the Diagnosis and Treatment Monitoring of Parkinson's Disease'. Annu Rev Biomed Eng. 2019;21: 111–143. doi:10.1146/annurev-bioeng-062117-121036

[8] Espay AJ, Bonato P, Nahab FB, Maetzler W, Dean JM, Klucken J, et al. 'Technology in Parkinson's disease: Challenges and opportunities'. Mov Disord. 2016;31: 1272–1282. doi:10.1002/mds.26642

[9] Caballol Pons N, Ávila A, Planas Ballvé A, Prats A, Quispe P, Pérez Soriano S, et al. 'Utilidad del sensor STAT-ON para la Enfermedad de Parkinson en la práctica clínica diaria'. Accepted. LXXIII Reunión Anual Sociedad Española de Neurología. 2021.

[10] Bayés À, Samá A, Prats A, Pérez-López C, Crespo-Maraver M, Moreno JM, et al. A '"HOLTER" for Parkinson's disease: Validation of the ability to detect on-off states using the REMPARK system'. Gait & Posture. Elsevier; 2018;59: 1–6. doi:10.1016/j.gaitpost.2017.09.031

[11] Caballol N, Prats A, Quispe P, Ranchal M, Alcaine S, Fondevilla F, et al. 'Early detection of Parkinson's disease motor fluctuations with a wearable inertial sensor'. Movement Disorders. 2020. p. 35 (suppl.

1). Available: www.mdsabstracts.org/abstract/early-detection-of-parkinsons-disease-motor-fluctuations-with-a-wearable-inertial-sensor/

[12] Caballol N, Perez-Soriano A, Planas-Ballvé A, Ávila A, Quispe P, Bayes À. 'Improving the patient self-awareness of the first motor fluctuations in early Parkinson's disease with a wearable sensor'. Mov Disord 2022. Madrid: Movement Disorders Society; 2022. Available: https://www.mdsabstracts.org/abstract/improving-the-patient-self-awareness-of-the-first-motor-fluctuations-in-early-parkinsons-disease-with-a-wearable-sensor/

[13] Caballol N, Perez-Lopez C, Perez-Soriano A, PLanas-Ballvé A, Ávila A, Quispe P, et al. 'Exploring the morning akinesia in Parkinson's disease with the wearable sensor STAT-ON'. Mov Disord 2022. Madrid: Movement Disorders Society; p. 1. Available: https://www.mdsabstracts.org/abstract/exploring-the-morning-akinesia-in-parkinsons-disease-with-the-wearable-sensor-stat-on/

[14] Nonnekes J, Snijders AH, Nutt JG, Deuschl G, Giladi N, Bloem BR. 'Freezing of gait: a practical approach to management'. Lancet Neurol. 2015;14. doi:10.1016/S1474-4422(15)00041-1

[15] Pötter-Nerger M, Dutke J, Lezius S, Buhmann C, Schulz R, Gerloff C, et al. 'Serum neurofilament light chain and postural instability/gait difficulty (PIGD) subtypes of Parkinson's disease in the MARK-PD study'. J Neural Transm. 2022;129: 295–300. doi:10.1007/s00702-022-02464-x

[16] Leta V, Jenner P, Chaudhuri KR, Antonini A. 'Can therapeutic strategies prevent and manage dyskinesia in Parkinson's disease? An update'. Expert Opin Drug Saf. 2019;18: 1203–1218. doi:10.1080/14740338.2019.1681966

[17] Rodríguez-Molinero A, Pérez-López C, Samà A, Rodríguez-Martín D, Alcaine S, Mestre B, et al. 'Estimating dyskinesia severity in Parkinson's disease by using a waist-worn sensor: concurrent validity study'. Sci Rep. 2019;9: 13434. doi:10.1038/s41598-019-49798-3

[18] Santos-García D, Deus Fonticoba T, Suárez Castro E, Aneiros Díaz A, McAfee D, Catalán MJ, et al. 'Non-motor symptom burden is strongly correlated to motor complications in patients with Parkinson's disease'. Eur J Neurol. 2020;27: 1210–1223. doi:10.1111/ene.14221

[19] Stocchi F, Antonini A, Barone P, Tinazzi M, Zappia M, Onofrj M, et al. 'Early DEtection of wEaring off in Parkinson disease: The DEEP study'. Parkinsonism Relat Disord. Elsevier; 2014;20: 204–211. doi:10.1016/J.PARKRELDIS.2013.10.027

[20] Martínez-Fernández R, Schmitt E, Martinez-Martin P, Krack P. 'The hidden sister of motor fluctuations in Parkinson's disease: A review on

nonmotor fluctuations'. Mov Disord. 2016;31: 1080–1094. doi:10.1002/mds.26731

[21] Antonini A, Moro E, Godeiro C, Reichmann H. 'Medical and surgical management of advanced Parkinson's disease'. Mov Disord. 2018;33: 900–908. doi:10.1002/mds.27340

[22] Nijhuis FAP, Elwyn G, Bloem BR, Post B, Faber MJ. 'Improving shared decision-making in advanced Parkinson's disease: protocol of a mixed methods feasibility study'. Pilot Feasibility Stud. 2018;4: 94. doi:10.1186/s40814-018-0286-4

[23] Antonini A, Stoessl AJ, Kleinman LS, Skalicky AM, Marshall TS, Sail KR, et al. 'Developing consensus among movement disorder specialists on clinical indicators for identification and management of advanced Parkinson's disease: a multi-country Delphi-panel approach'. Curr Med Res Opin. 2018;34: 2063–2073. doi:10.1080/03007995.2018.1502165

[24] Santos-Garciá D, De Deus Fonticoba T, Suárez Castro E, Aneiros Diáz A, McAfee D. '5-2-1 Criteria: A Simple Screening Tool for Identifying Advanced PD Patients Who Need an Optimization of Parkinson's Treatment'. Parkinsons Dis. 2020;2020. doi:10.1155/2020/7537924

[25] Malaty IA, Martinez-Martin P, Chaudhuri KR, Odin P, Skorvanek M, Jimenez-Shahed J, et al. 'Does the 5–2-1 criteria identify patients with advanced Parkinson's disease? Real-world screening accuracy and burden of 5–2-1-positive patients in 7 countries'. BMC Neurol. 2022;22: 35. doi:10.1186/s12883-022-02560-1

[26] Martinez-Martin P, Kulisevsky J, Mir P, Tolosa E, García-Delgado P, Luquin M-R. 'Validation of a simple screening tool for early diagnosis of advanced Parkinson's disease in daily practice: the CDEPA questionnaire'. npj Park Dis. 2018;4: 20. doi:10.1038/s41531-018-0056-2

[27] Antonini A, Odin P, Schmidt P, Cubillos F, Standaert DG, Henriksen T, et al. 'Validation and clinical value of the MANAGE-PD tool: A clinician-reported tool to identify Parkinson's disease patients inadequately controlled on oral medications'. Parkinsonism Relat Disord. 2021;92: 59–66. doi:10.1016/j.parkreldis.2021.10.009

[28] Moes HR, Buskens E, van Laar T. Letter to the editor, 'Validation and clinical value of the MANAGE-PD tool: A clinician-reported tool to identify Parkinson's disease patients inadequately controlled on oral medications.' Parkinsonism Relat Disord. 2022;97: 99–100. doi:10.1016/j.parkreldis.2022.03.014

[29] Rodríguez-Martín D, Cabestany J, Pérez-López C, Pie M, Calvet J, Samà A, et al. 'A New Paradigm in Parkinson's Disease Evaluation

With Wearable Medical Devices: A Review of STAT-ON™'. Front Neurol. 2022;13. doi:10.3389/fneur.2022.912343

[30] Rodríguez-Molinero A, Pérez-López C, Samá A, De Mingo E, Rodríguez-Martín D, Hernández-Vara J, et al. 'A kinematic sensor and algorithm to detect motor fluctuations in Parkinson disease: Validation study under real conditions of use'. J Med Internet Res. 2018;20. doi:10.2196/rehab.8335

[31] Pérez-López C, Hernández-Vara J, Caballol N, Bayes À, Buongiorno M, Lopez-Ariztegui N, et al. 'Comparison of the Results of a Parkinson's Holter Monitor With Patient Diaries, in Real Conditions of Use: A Subanalysis of the MoMoPa-EC Clinical Trial'. Front Neurol. 2022;13. doi:10.3389/fneur.2022.835249

[32] Santos García D, López Ariztegui N, Cubo E, Vinagre Aragón A, García-Ramos R, Borrué C, et al. 'Clinical utility of a personalized and long-term monitoring device for Parkinson's disease in a real clinical practice setting: An expert opinion survey on STAT-ON™'. Neurología. 2020; doi:10.1016/j.nrl.2020.10.013

[33] Barrachina-Fernández M, Maitín AM, Sánchez-Ávila C, Romero JP. 'Wearable Technology to Detect Motor Fluctuations in Parkinson's Disease Patients: Current State and Challenges'. Sensors. 2021;21: 4188. doi:10.3390/s21124188

[34] Storch A, Schneider CB, Wolz M, Sturwald Y, Nebe A, Odin P, et al. 'Nonmotor fluctuations in Parkinson disease: Severity and correlation with motor complications'. Neurology. 2013;80: 800–809. doi:10.1212/WNL.0b013e318285c0ed

[35] Honig H, Antonini A, Martinez-Martin P, Forgacs I, Faye GC, Fox T, et al. 'Intrajejunal levodopa infusion in Parkinson's disease: A pilot multicenter study of effects on nonmotor symptoms and quality of life'. Mov Disord. 2009;24: 1468–1474. doi:10.1002/mds.22596

[36] Kurtis MM, Rajah T, Delgado LF, Dafsari HS. 'The effect of deep brain stimulation on the non-motor symptoms of Parkinson's disease: a critical review of the current evidence'. npj Park Dis. 2017;3: 16024. doi:10.1038/npjparkd.2016.24

[37] Bougea A, Palkopoulou M, Pantinaki S, Antonoglou A, Efthymiopoulou F. 'Validation of a real-time monitoring system to detect motor symptoms in patients with Parkinson's Disease treated with Levodopa Carbidopa Intestinal Gel'. International Congress of Parkinson's Disease and Movement Disorders. 2021. Available: https://virtual.mdscongress.org/posters/28135960/Validation-of-a---real--time-monitoring-system--to-detect--motor-

symptoms-in-patients-with-Parkinsons-disease-treated-with-Levodopa-Carbidopa-Intestinal-Gel

[38] Herreros Rodriguez J, Esquivel López A, Romero Muñoz J, Llaguno Velasco M. 'Use of a wearable medical device for LCIG tube adjustment: a user case'. Movement Disorders, editor. 2022 International Congress of Parkinson's Disease and Movement Disorders. 2022. p. 37-supp 1.

[39] Barrios López JM, Ruiz Fernández E, Triguero Cueva L, Madrid Navarro C, Pérez Navarro MJ, Jouma Katati M, et al. 'Simultaneous recording in Parkinson's disease with STAT-ON™ and subthalamic local field potentials (Percept™)'. 8th Congress of the European Academy of Neurology - Europe 2022. 2022.

[40] Perrote F, Zeppa G, Coca H, Figueroa S, de Battista JC. 'Evaluación de un sistema de sensores inerciales externos tipo Holter en pacientes con enfermedad de Parkinson en Argentina'. Neurol Argentina. 2021;13: 153–158. doi:10.1016/j.neuarg.2021.05.006

[41] Sense4Care. STAT-ON Holter 2021 Available: www.statonholter.com

[42] Gatekeeper Project. Parkinson's Disease Treatment. [Internet]. [cited 13 Feb 2023]. Available: https://www.gatekeeper-project.eu/parkinsons-disease-treatment-dss/

[43] Schrag A, Quinn N. 'Dyskinesias and motor fluctuations in Parkinson's disease'. Brain. 2000;123: 2297–2305. doi:10.1093/brain/123.11.2297

[44] Seppi K, Ray Chaudhuri K, Coelho M, Fox SH, Katzenschlager R, Perez Lloret S, et al. 'Update on treatments for nonmotor symptoms of Parkinson's disease—an evidence-based medicine review'. Mov Disord. 2019;34: 180–198. doi:10.1002/mds.27602

[45] Channa A, Popescu N, Ciobanu V. 'Wearable Solutions for Patients with Parkinson's Disease and Neurocognitive Disorder: A Systematic Review'. Sensors. 2020;20: 2713. doi:10.3390/s20092713

[46] Rodriguez-Molinero A, Hernández-Vara J, Miñarro A, Martinez-Castrillo JC, Pérez-López C, Bayes À, et al. 'Multicentre, randomised, single- blind, parallel group trial to compare the effectiveness of a Holter for Parkinson's symptoms against other clinical monitoring methods: study protocol'. BMJ Open. 2021;11: 1–9. doi:10.1136/bmjopen-2020-045272

[47] Pagano G, Boess FG, Taylor KI, Ricci B, Mollenhauer B, Poewe W, et al. 'A Phase II Study to Evaluate the Safety and Efficacy of Prasinezumab in Early Parkinson's Disease (PASADENA): Rationale, Design, and Baseline Data'. Front Neurol. 2021;12. doi:10.3389/fneur.2021.705407

[48] Sánchez-Ferro Á, Maetzler W. 'Advances in sensor and wearable technologies for Parkinson's disease.' Mov Disord. 2016;31: 1257–1257. doi:10.1002/mds.26746

[49] Godinho C, Domingos J, Cunha G, Santos AT, Fernandes RM, Abreu D, et al. 'A systematic review of the characteristics and validity of monitoring technologies to assess Parkinson's disease'. J Neuroeng Rehabil. 2016;13: 24. doi:10.1186/s12984-016-0136-7

[50] Rovini E, Maremmani C, Cavallo F. 'How Wearable Sensors Can Support Parkinson's Disease Diagnosis and Treatment: A Systematic Review'. Front Neurosci. 2017;11. doi:10.3389/fnins.2017.00555

[51] Sica M, Tedesco S, Crowe C, Kenny L, Moore K, Timmons S, et al. 'Continuous home monitoring of Parkinson's disease using inertial sensors: A systematic review'. Barbieri FA, editor. PLoS One. 2021;16: e0246528. doi:10.1371/journal.pone.0246528

[52] Hadley AJ, Riley DE, Heldman DA. 'Real-World Evidence for a Smartwatch-Based Parkinson's Motor Assessment App for Patients Undergoing Therapy Changes'. Digit Biomarkers. 2021;5: 206–215. doi:10.1159/000518571

[53] Tanguy A, Jönsson L, Ishihara L. 'Inventory of real world data sources in Parkinson's disease'. BMC Neurol. 2017;17: 213. doi:10.1186/s12883-017-0985-0

[54] van Wamelen DJ, Sringean J, Trivedi D, Carroll CB, Schrag AE, Odin P, et al. 'Digital health technology for non-motor symptoms in people with Parkinson's disease: Futile or future?' Parkinsonism Relat Disord. 2021;89: 186–194. doi:10.1016/j.parkreldis.2021.07.032

[55] Heinzel S, Lerche S, Maetzler W, Berg D. 'Global, Yet Incomplete Overview of Cohort Studies in Parkinson's disease'. J Parkinsons Dis. 2017;7: 423–432. doi:10.3233/JPD-171100

[56] Kleiner G, Fernandez HH, Chou KL, Fasano A, Duque KR, Hengartner D, et al. 'Non-Motor Fluctuations in Parkinson's Disease: Validation of the Non-Motor Fluctuation Assessment Questionnaire'. Mov Disord. 2021;36: 1392–1400. doi:10.1002/mds.28507

[57] Chirra M, Marsili L, Wattley L, Sokol LL, Keeling E, Maule S, et al. 'Telemedicine in Neurological Disorders: Opportunities and Challenges'. Telemed e-Health. 2019;25: 541–550. doi:10.1089/tmj.2018.0101

[58] Heldman DA, Giuffrida JP, Cubo E. 'Wearable Sensors for Advanced Therapy Referral in Parkinson's Disease'. J Parkinsons Dis. 2016;6: 631–638. doi:10.3233/JPD-160830

[59] Willows T, Dizdar N, Nyholm D, Widner H, Grenholm P, Schmiauke U, et al. 'Initiation of Levodopa-Carbidopa Intestinal Gel Infusion Using Telemedicine (Video Communication System) Facilitates Efficient and Well-Accepted Home Titration in Patients with Advanced Parkinson's Disease'. J Parkinsons Dis. 2017;7: 719–728. doi:10.3233/JPD-161048

[60] Cubo E, Garcia-Bustillo A, Arnaiz-Gonzalez A, Ramirez-Sanz JM, Garrido-Labrador JL, Valiñas F, et al. 'Adopting a multidisciplinary telemedicine intervention for fall prevention in Parkinson's disease. Protocol for a longitudinal, randomized clinical trial'. PLoS One. 2021;16: e0260889. doi:10.1371/journal.pone.0260889

[61] Grahn F. 'Evaluation of Two Commercial Sensor Systems for Monitoring Parkinsonism and Their Possible Influence on Management of Parkinson's Disease'. Institute of Neuroscience and Physiology Sahlgrenska Academy University of Gothenburg. 2022. Available: http://hdl.handle.net/2077/70780

8

New Open Scenarios for STAT-ON™: The Business Perspective

Joan Calvet and Chiara Capra

Sense4Care SL – Cornellà de Llobregat, Spain

Email: (joan.calvet) (chiara.capra)@sense4care.com

Abstract

This chapter provides a general overview on the open new business perspective when using the appropriate technology to implement new eHealth services and consultations. By focusing on Parkinson's disease only, this chapter will encompass the incidence of PD, the importance of patient-centered care, how it can benefit from technologies, the market size, and opportunities at the healthcare ecosystem level. Finally, different use cases and STAT-ON™ applications are presented where relevant. Given the clear lack in the clinical evaluation of PD, we conclude that a global claim for technologies is recognized. Furthermore, given its extended scientific backup, including key validation studies, STAT-ON™ can be considered as the new gold standard for PD evaluation.

8.1 Introduction: A General Overview

The incidence of Parkinson's disease (PD) in our society is significant and is growing exponentially. About 8.5 million people worldwide have been diagnosed according to WHO, a neurodegenerative disorder with disabling effects on its sufferers and with no cure. Only in Europe, there are 1.2M patients, according to Parkinson's Europe Association.

The prevalence of the disease ranges from 41 people per 100,000 in the fourth decade of life to more than 1900 people per 100,000 among those 80 and older [1]. It is expected to double in 2040 or even triple, assuming other factors than aging [2]. PD is the second neurodegenerative disease around the world after Alzheimer's disease. The incidence of the disease has an estimated 4% of people diagnosed before the age of 50 [3], and PD has a high impact on life duration expectations and Quality of Life (QoL) for all the persons affected by it.

In daily clinical practice, healthcare professionals, patients, and caregivers have a hard time to make a complete and objective clinical assessment of PD patients. Due to high costs and major time spent, usually, patients are examined once or twice a year with a relatively brief clinical evaluation. When treating PD, neurologists use a methodology to evaluate the disease progression mainly based on a report filled out during the patient's visit. Also, patients tend to show up to the visit post to medication intake, which leads to difficulties presenting real symptoms of the OFF state of the patient while in front of the doctor. Moreover, the "white coat effect" and the "Hawthorne effect," intended as the behavioral change due to the awareness of the patient of being evaluated, affect the severity of the symptoms presented during the doctor's visit.

Furthermore, for remote symptomatology monitoring, healthcare professionals must rely on patients' diaries, which patients often have reduced compliance to, and major recall bias. In addition, the current procedure often leads to a subjective evaluation and a lack of information in the doctor's office to properly evaluate the PD patient. Thus, there is a big claim for more objective measurements in order to provide more home-environment information and daily life symptoms and achieve a more accurate diagnosis and follow-up, leading to more efficient therapy management. In this regard, **STAT-ON™ answers to these issues by overcoming the subjective and well-scientifically demonstrated difficulties with questionnaires for PD patients. STAT-ON™, in fact, provides objective information about the severity and distribution of PD motor symptoms and their fluctuations in daily life, allowing for unbiased monitoring of the patient.**

Motor fluctuations are the most perturbing symptoms, according to patients. According to the DEEP study from Stocchi et al. [4], there is an infra-diagnosis of the first fluctuations leading to providing an inappropriate therapy already from the early symptoms' detection, directly impacting QoL in further years. Given the strong symptomatic ON/OFF fluctuations and the failure to accurately track the progression of the disease by the existing standard of care, resulting in poor QoL and higher dependence on the patient.

A worse QoL impacts economics, with more hospitalizations, productivity loss, and additional care problems. PD turns into a total social cost of €13.9B per year, meaning €11,6K for patient/year on average, including direct medical costs and loss of productivity of both sufferers and caregivers in Europe [5].

It is important to highlight that in PD, the main cost is not associated to drugs (4.4–20% depending on the analysis) but to the support and nursing that lead to €11Bn in the EU, €9200 per patient as the average of total cost of care (TCOC). However, these costs can be lowered substantially with the use of a correct technological support because STAT-ON™ can alert the physician about the real state of the patient in terms of motor state, focusing more concretely on aspects such as gait disturbances or freezing of gait (FoG), which leads to falls. Recently, it has been demonstrated that STAT-ON™ is able to detect early fluctuations and dyskinesia [6–8], leading to an appropriate titration of the patient and an accurate early therapy prescription.

8.2 The Use of Technology for a Patient-Centered Care

The recent COVID-19 pandemic has highlighted using technology as key for patients' remote monitoring and improving treatment and diagnostic options. The pandemic has led, in fact, to a striking evolution in the use of telehealth, intended as the variety of technologies and services to offer patient care and improve the healthcare delivery system as a whole. The scientific community has, in fact, claimed the major need for technologies for a better and more efficient clinical practice.

Within the neurodegenerative disorders field, several neurologists state that due to the rise of the burden of neurodegenerative disorders, the healthcare systems will suffer from a deficiency of basic healthcare services in the next few years. Given this scenario, a reliable tool to detect a patient's motor state objectively and remotely is key to delivering better neurodegenerative disease management. **Therefore, the appropriate technology will be essential to transform the classical PD evaluation, moving forward to a new paradigm in clinical practice.** In this regard, the STAT-ON™ Holter postulates as the best technology to achieve this goal, given its characteristics, clinical validation, and scientific endorsement.

In PD, there is a clear need for new methods to permanently track PD symptoms and improve PD patient's state to properly care for them diminishing nursing and medical costs. STAT-ON™ is a clear empowering tool for patients, which allows to track and send remotely reliable and objective information in home environments, which physicians would otherwise obtain incorrectly in ambulatory conditions.

> As a consequence, the integration of STAT-ON™ in telehealth and telemedicine will allow for better and more efficient remote monitoring, improved therapy, improved assessment, and improved disease management as a whole, in clinics, home healthcare, as well as in both public and private hospitals.

In this regard, in the past decade, the concept of healthcare has moved from a clinician-centered vision to patient-centered care. The latter refers as the practice in which patients actively participate in their own medical treatment in close collaboration with their healthcare professionals for better patients' outcomes. Among its key pillars is access to medical information and education on specific diseases, as well as easing access to care. STAT-ON™'s reports, comprising weekly graphs, reporting patients' symptomatology patterns, and specific tailored variables on patients' gait with personal set thresholds, allow the patient better to understand their key symptoms and fluctuations throughout the day. In this way, the healthcare professional can educate their patients about their disease stage and progression, making them feel part of their therapy and therapeutic adjustments.

Improvements in therapy adjustments, in fact, lead to improved patients' QoL, and, therefore, to improved healthcare professional–patient relationship. Moreover, STAT-ON™, when integrated into telehealth and telemedicine, allows patients to have better access to more personalized care through technology. STAT-ON™, in fact, enables continuing access to the improvement of patients' care thanks to its data and tailored analytics. Furthermore, the usability of technology, especially for body-worn devices, is essential, and it is crucial to ease the setting of the sensors to maintain the adherence of the patient to the technology. Usability needs to be tailored to the patient and can be considered under different perspectives: ease of use, number of sensors, and part of the body where the sensor or sensors are located. It is well known that, if the system implies wearing more sensors, the precision may be higher for a more variety of movements, but user satisfaction and experience drop drastically. In this regard, **STAT-ON™ has been demonstrated to score high on usability** thanks to its comfortable belt which allows the patient to wear it in home environments.

Thus, it can be claimed that telehealth will provide key benefits both at the individual level and the healthcare system level. Major benefits have been identified, among which:

- programmed clinical monitorizations,

- annual cost-saving visits due to accurate diagnosis,
- follow-up compared to routine care only,
- technological innovation introduction at a large scale in different healthcare systems,
- greater accessibility to large-scale clinical data at a governmental level,
- better management and control of the disease in remote and rural areas,
- improved patients' and clinical experience thanks to better decision-making outcomes,
- reduction of disease burden for the healthcare system and the community.

8.3 Market Size and Impact

STAT-ON™ can be considered as part of four different but connected, rising markets (medical wearables devices, telemedicine, Parkinson's disease, and artificial intelligence). As mentioned in the previous chapters, STAT-ON™ is a wearable medical device aimed at monitoring motor symptoms of PD. It is based on artificial intelligence algorithms and could be used to monitor a patient remotely.

The market of medical technologies, more concretely of wearable medical devices, is a rising market that enables physicians to get voluminous patient data in real environments, allowing them to perform more accurate evaluations. The Compound Annual Growth Rate (CAGR) is estimated to be 20.5% in 5 years, and the expected market size in 2026 is $46Bn.

The COVID-19 pandemic, among other socioeconomic and technological factors, has led to accelerated adoption of new technologies to support healthcare professionals and even replace some of the methodologies used in the last decades. In the case of Parkinson's disease, patients have decreased the number of visits but worsening their symptoms at not being well treated. This, along with the advance of Internet of Things - IoT devices, telehealth apps, virtual hospitals, made healthcare API market and telemedicine growth faster in the last years. The global market size in 2030 is estimated at $310Bn with a CAGR of 3.56% from 2020 to 2028.

Parkinson's disease global drug market forecast in 2028 is expected to be $12.3Bn with a CAGR of 12.3 over 2022 and 2028. It is a rising market, as patients are forecasted to double in 2040 compared to numbers in 2022. According to many presentations in the Movement Disorders Society event MDS2021, there is a clear need to detect patients earlier in each one of the stages of Parkinson's disease in order to provide them with a correct and tailored therapy. Pharmaceutical companies are competing between them to situate their drug solution in a specific stage of Parkinson's disease. However, the main issue comes in the evaluation of the patient, which continues to be problematic, subjective, and with few information about the real state of the patient. STAT-ON™ will help the physicians in the decision-making process and will allow to detect a patient earlier for a concrete treatment or therapy improvement (as an example, see a collection of real cases in Chapter 6).

Finally, artificial intelligence is a growing market. With the advent of machine learning techniques, deep learning, and big data, artificial intelligence will gain more importance in the coming years. The CAGR is 38.1% from 2022 to 2030, and the market size, although there are many different conclusions and results, it is estimated to be $1,591Bn by 2030[1].

The global medical devices reimbursement market was at $427Bn in 2021 and is expected to reach over $860Bn by 2030, with a Compound Annual Growth Rate (CAGR) estimated at 8.1% in this period.

8.4 STAT-ON™: The New Gold Standard

When discussing the competitive landscape, it is convenient to consider the existing medical devices, among which is STAT-ON™. Other technologies, not certified as medical devices, claim that they are valid solutions, although only the certification process can guarantee the characteristics of accuracy, reliability, and safety for the patient.

Many of the existing competitive solutions to STAT-ON™ exhibit inferior characteristics, and the main reason is their location on the patient's body when they are wearing it (many of them are placed on the wrist), which means that they cannot correctly detect, with the required reliability, the symptoms that they are supposed to detect (bradykinesia, FoG, etc.). Other solutions are in different stages of technological development; however, they have not yet gone through the certification process.

A clear advantage of STAT-ON™ is its easiness of use, combined with the capacity to register with a single unique device the motor state in home-environment conditions with clinically validated advanced machine learning algorithms. As a result, the system is a single certified medical device, based on inertial technology and worn at the waist, from where PD

---

[1] Market reports https://www.precedenceresearch.com

motor symptoms such as bradykinesia, gait disturbances, Freezing of Gait, or dyskinesia, can be very well detected, characterized, and registered.

The manufacturing company (Sense4Care SL) aims to establish STAT-ON™ and all its derivate devices as a gold standard for monitoring Parkinson's disease in home environments, directly impacting the evaluation of PD patients in clinical practice and during the execution of clinical trials.

The following text will present and discuss four different use cases, to open the mind to new services, business approaches, empowerment of the patients, and the establishment of new relationships between patients and their neurologists.

Use case 1: The case for early and advanced PD detection
According to some discussion with different players in the field (different companies, business developers, and neurologists), one relevant conclusion is that STAT-ON™ is a considerable tool for neurologists to detect PD at each stage, and however, early PD and Advanced PD detection have been demonstrated to be the most applicable stages, and pharmaceutical and MedDev companies can benefit from STAT-ON™ to detect patients earlier.

Early PD detection can be challenging as motor fluctuation recognition is not always clear to the patients, and therefore, they cannot explain themselves properly in the clinical consultation. However, STAT-ON™ has been recently recognized as a useful tool to detect motor fluctuations, even if patients were not aware of their symptoms or did not report any kind of symptom [6, 7]. Similarly, in another study, morning akinetic patients were detected by using STAT-ON™ by analyzing the gait fluidity only [9]. Thus, as mentioned in the previous chapter, although further evidence needs to be generated, STAT-ON™ could be useful for early detection of predictable fluctuations (morning akinesia; wearing-off) and double-checking whether the patient is actually having these fluctuations or not. Moreover, as the disease progresses, STAT-ON™ can be used to monitor these fluctuations, check daily motor patterns, and predict the patient's ON and OFF states during the day. **Indeed, in the early PD stage, STAT-ON™ can be useful to provide the right and most tailored therapy possible to the patient or to make a tailored therapeutic adjustment.**

Like detecting motor fluctuations in early PD stages, STAT-ON™ can also be useful for detecting patients with advanced PD symptoms (APD) needing second-line therapy. In the DISCREPA study [10], the authors found that around 30% of APD are not well diagnosed as APD and continue taking conventional drugs that do not allow an acceptable QoL. When a patient has rated H and Y III (moderate-advanced), they reach a point where conventional therapies are ineffective, and the only way to improve QoL is to use advanced therapies.

STAT-ON™ can monitor patients from stages I to IV. Patients with H and Y stage IV and V are considered APD. Patients from H and Y = IV represent almost 20% of patients with PD, but according to the aforementioned study, 30% are not well-diagnosed. In other words, in Spain, there are approximately 20,000 patients that can be monitored with STAT-ON™ that are in APD. From this number of patients and according to this study, there are around 6000 patients that are not well diagnosed and need a second-line therapy such as DBS, apomorphine, or duodopa infusion pumps. In France's case, 50,000 patients with a H and Y of 4 and 15,000 patients would need APD therapy. In the case of the UK, numbers are very similar to Spain.

STAT-ON™ can provide clear and objective information to detect APD patients, according to a Spanish study performed in 27 hospitals in Spain. A total of 81.5% of neurologists think that it is a very useful tool to detect APD patients, and thus, that need APD therapies by observing the time in OFF and the dyskinesia suffered by the patient in an objective way [11].

On the other hand, primary care centers (PCC) are centers where a PD patient is usually attended only in case there is very clear evidence of the condition of the patient and the impression of the generic neurologist that the patient does not respond to medication, then the patient is derived to a second or third level hospital. This process can be advanced by providing clear information (STAT-ON™ can provide it) to these nonspecialist health professionals. **The patients do not have to get to situations where their conditions are harmful, and their QoL is extremely low**. Pharmaceutical companies can be the main beneficiaries of this correct evaluation of the patients. The new evaluation and detection service, using STAT-ON™, will approach the patient to an advanced therapy when required and on time.

Use case 2: Better attention in public hospitals

The main issue of a public hospital is the saturation of the health system and difficulties in managing the patients' visits. Thus, one of the main aims is to decrease this saturation, the number of visits, and the time spent out of the visits by doing reports or therapy adjustments.

An example was given by a neurologist in Barcelona, who declared that he could spend 2 hours per patient to correctly adjust a DBS system based on the information received from diaries and questionnaires. In the COVID-19 scenario, this is one of the main objectives (decrease of the saturation of the services and the minimization of the physical visits to the hospital).

Some neurologists that have widely used STAT-ON™ stand with the preliminary conclusion that **although STAT-ON™ does not reduce the time of visit, a significant improvement in the visit quality can be noted**. When the neurologist gains some experience with the use of STAT-ON™, they claim that the sensor could speed up the visit. They also state that providing the correct therapy might reduce the number of visits per patient as the therapy is correct, and patient does not need to come with such frequency to correct the therapy provided.

In this term, a new paradigm needs to be set in the public system. The sensor can be sent to the PD patient's home, and then remote monitoring can be performed. The neurologists can call the patient to perform a follow-up. This process would prevent the patient from mobilizing to the hospital. Also, management can be performed remotely and with very good and precise objective information.

This new paradigm contributes innovation to hospitals benefiting from funded projects and distinction as reference hospitals. Rigorous conclusions must be analyzed in these studies that are being performed to set up a profitable business model. **Quality of life improvement of patients is not the unique and main goal from hospitals as stated some interviewed neurologists, and costs and time are of great interest for hospital managers.**

Use case 3: Clinical trials
A clinical trial is a very important part for the development of new treatments, performed by international research groups or by the principal pharmaceutical companies.

In a clinical trial, several control processes must be taken from the analysis of patients. For example, complex but subjective questionnaires or diaries must be obtained after filled-in with a costly supervision process for the quality of these data. In a clinical trial, which is normally funded by public entities or pharmaceutical companies, there are some key points to minimize: **the time and cost of these processes**.

The main target for reducing time and cost is all the activities performed over a PD patient out of the visit. For example, interpreting diaries, controlling the process (calls or visits to patient's home), and checking if the diary was correct or should be repeated. All these processes are cumbersome and, as reported by the PI of a clinical trial performed with STAT-ON™, diaries must be repeated several times, and it is difficult to have clear and reliable annotations. It is well known that not many patients can efficiently fill their diaries properly. Thus, the sensor can cover all patients, and there is no need to precisely control patients to fill diaries properly. The Madrid's

Parkinson's Association, in Spain, has recognized that STAT-ON™ can perfectly substitute the diary process as standardize all the metrics filled in a diary, and the information is always objective and real. Also, in a study conducted by Santos et al. with 27 movement disorders experts, 70.3% stated that STAT-ON™ was better than a diary only by having used once [11]. Diaries have the issue of reduced compliance and recall bias.

The UPDRS is the most common questionnaire used in clinical practice. It takes about 10 minutes to be filled out. The questionnaire is quite subjective, but most neurological community has standardized and accepted it. Adding the time for interpretation, digitalizing it, and making decisions, take time to the neurologist, approximately 1 hour per patient. The sensor offers most of the information provided by UPDRS on motor symptoms, reducing the time of filling and digitalizing it. Also, it provides objective information.

The time to understand the STAT-ON™ report is about 15 minutes. For clinical trials, the longer version of UPDRS is used (the UPDRS-MDS), which can take about 30 minutes to complete the questionnaire. The use of STAT-ON™ makes it not necessary to spend time filling in diaries and reduces the time spent interpreting the results of the analysis.

The possible outcomes of this process are: it is not necessary the participation of so many health professionals in a PD study, reducing the time of visits, the time of interpretation, and digitalization of diaries. However, the idea is to increase the PD patients to perform studies with more data consistency. The same staff could perform approximately the double of databases by using a sensor. Moreover, all pilots would gather objective data from the sensor and also in real-life conditions, being very productive to a pharma company as they could show their medication used in real-life conditions.

Use case 4: Improvement of the service in private clinics and home-healthcare centers

These specific centers normally integrate a service with different departments (neurology, physiotherapy, psychology, etc.), offering a global health and care service to PD patients. These centers compete with others to attract patients to their services, by offering a better service which can be translated as a significant improvement in QoL.

According to the personal experience of two neurologists working in different private centers, there are two key points to analyze: **the reimbursement policy, the quality of the visit, and the service offered to the patient**.

It has already been shown that STAT-ON™ improves the quality of visit, offering a better treatment and enhancing QoL. However, additional considerations should be made for adding new quality services, based on the use of the technology, and investigating how these initiatives would be made profitable as part of the day-by-day clinical praxis.

This case can be generalized to home healthcare (HHC) in many countries (France, in particular), where different HHC service providers have been detected, and a business model based on the massive use of STAT-ON™ must be found.

The center can empower the patient by offering him an extra service with STAT-ON™. They can also provide a STAT-ON™ to the patient, who can pay monthly rent or fee.

STAT-ON™ provides different advantages that a private center can benefit, as an improvement in therapy adjustment, which means improvement in QoL, satisfying the patient, and increasing the quality of the relationship between the neurologist and the patient. Also, the center can attend to more patients as professionals will have digital tools to evaluate objectively without doing classical evaluations such as diaries or questionnaires.

According to the previous statement, the patient can also have more visits if they note a decrease in the therapeutic effect. They will not have to wait 6–8 months. Finally, it adds innovation and added value to the offered service and offers the possible implementation of the automatic generation of alarms when some previously established thresholds are reached by specific parameters or combinations of parameters.

8.5 Conclusions

There are several drawbacks in the current evaluation of PD, and various studies claim the use of new technologies to provide useful, objective, and clear information of PD patients in home environments. Several patients and also neurologists enter a continuous loop where the therapy is not well adjusted, the QoL of the patient decreases, and they need another visit. The use of new technologies would offer this information, and thus, therapies could be more accurate.

STAT-ON™ is a medical device, class IIa, that acts as a real Holter for the motor symptoms of PD patients, with a high rate of usability. There are several advantages to STAT-ON™ use, and it should be a cost-effective solution for many stakeholders.

On the one hand, companies that manufacture second-line therapies can benefit from the fact that STAT-ON™ can be used to detect advanced patients. Given that there is a high rate of patients not well-diagnosed with APD, detecting these patients would lead to an increase in the use of these therapies. This fact involves two main advantages: an increase in QoL for the patient and an economic benefit for the pharma that manufactures the advanced therapy. The same happens with first fluctuations and the need to prescribe dopaminergic inhibitors at the right moment. STAT-ON™ has demonstrated to clearly identify first motor fluctuations and the need to complement the levodopa-based therapies for reducing the time in OFF.

Also, the STAT-ON™ can help to manage better clinical trials by providing faster, more reliable, and objective information of the state of the patient, reducing the cost drastically, and speeding up the clinical trials.

In the field of hospitals, STAT-ON™ is also useful for decreasing the number of face-to-face visits (even eliminating them by remote monitoring with the sensor and a telephone call) and improving the quality of the visit by providing more accurate therapy to the patient. Innovation is also a quality of excellence to consider, as well as the possibility of managing the visits remotely given a pandemic scenario as it was the COVID-19.

Finally, several advantages have been presented in home-healthcare centers and private clinics, given that STAT-ON™ enhances the expertise and aptitudes of the center against other competitors and attracts more customers.

In conclusion, **STAT-ON™ has been presented as a perfect tool for health professionals, as an instrument that can be used to improve QoL of patients (benefiting the health systems by reducing medical care, and hospitalization admissions), and as an attractive device for pharmaceutical companies.**

References

[1] Cacabelos R. 'Parkinson's Disease: From Pathogenesis to Pharmacogenomics'. Int J Mol Sci. 2017;18: 551. doi:10.3390/ijms18030551

[2] Dorsey ER, Sherer T, Okun MS, Bloem BR. 'The Emerging Evidence of the Parkinson Pandemic'. Brundin P, Langston JW, Bloem BR, editors. J Parkinsons Dis. 2018;8: S3–S8. doi:10.3233/JPD-181474

[3] van den Eeden SK. 'Incidence of Parkinson's Disease: Variation by Age, Gender, and Race/Ethnicity'. Am J Epidemiol. 2003;157: 1015–1022. doi:10.1093/aje/kwg068

[4] Stocchi F, Antonini A, Barone P, Tinazzi M, Zappia M, Onofrj M, et al. 'Early DEtection of wEaring off in Parkinson disease: The DEEP study'. Parkinsonism Relat Disord. 2014;20: 204–211. doi:10.1016/J.PARKRELDIS.2013.10.027

[5] Gustavsson A, Svensson M, Jacobi F, Allgulander C, Alonso J, Beghi E, et al. 'Cost of disorders of the brain in Europe 2010'. European

Neuropsychopharmacology. 2011;21: 718–779. doi:10.1016/j.euroneuro.2011.08.008

[6] Caballol N, Prats A, Quispe P, Ranchal M, Alcaine S, Fondevilla F, et al. 'Early detection of Parkinson's disease motor fluctuations with a wearable inertial sensor'. Movement Disorders. 2020. p. 35 (suppl. 1). Available: www.mdsabstracts.org/abstract/early-detection-of-parkinsons-disease-motor-fluctuations-with-a-wearable-inertial-sensor/

[7] Caballol N, Bayés À, Prats A, Martín-Baranera M, Quispe P. 'Feasibility of a wearable inertial sensor to assess motor complications and treatment in Parkinson's disease'. PLoS One. 2023.

[8] Caballol N, Perez-Soriano A, Planas-Ballvé A, Ávila A, Quispe P, Bayes À. 'Improving the patient self-awareness of the first motor fluctuations in early Parkinson's disease with a wearable sensor'. Mov Disord 2022. Madrid: Movement Disorders Society; 2022. Available: https://www.mdsabstracts.org/abstract/improving-the-patient-self-awareness-of-the-first-motor-fluctuations-in-early-parkinsons-disease-with-a-wearable-sensor/

[9] Caballol N, Perez-Lopez C, Perez-Soriano A, PLanas-Ballvé A, Ávila A, Quispe P, et al. 'Exploring the morning akinesia in Parkinson's disease with the wearable sensor STAT-ON™'. Mov Disord 2022. Madrid: Movement Disorders Society; p. 1. Available: https://www.mdsabstracts.org/abstract/exploring-the-morning-akinesia-in-parkinsons-disease-with-the-wearable-sensor-stat-on/

[10] Ávila A, Pastor P, Planellas L, Gil-Villar MP, Hernández-Vara J, Fernández-Dorado A. 'DISCREPA study: Treatment of advanced Parkinson's disease and use of second-line treatments in Catalonia'. Rev Neurol. 2021;72: 1–8. doi:10.33588/RN.7201.2020181

[11] Santos García D, López Ariztegui N, Cubo E, Vinagre Aragón A, García-Ramos R, Borrué C, et al. 'Clinical utility of a personalized and long-term monitoring device for Parkinson's disease in a real clinical practice setting: An expert opinion survey on STAT-ON™'. Neurología. 2020. doi:10.1016/j.nrl.2020.10.013

Index

A
Accelerometer 40, 43, 45, 51, 53–54, 56, 60–63

D
Device-aided therapy 175, 179–181, 183–184, 213–216, 219

E
Early detection 113, 153, 233
Enclosure 38, 47, 52, 63–66, 68–69, 91, 97

F
Fall 10, 23, 51, 133, 136, 145–147
Firmware 37, 42, 53, 55, 62–63, 101–102, 111
Freezing of gait 2–3, 23, 28, 131, 154, 156, 162–163, 167, 173, 180, 186, 188, 198, 208, 210, 214, 229, 233
Future scenarios 218

G
Gold standard 227, 232–233

H
Hardware 39, 42, 62–63, 76, 92, 112

I
Interface 22–23, 29, 40, 42, 44, 54, 60–61, 76, 95, 117–120, 124–125, 145, 149

L
Labeling 37, 72–73, 88, 90, 92, 95, 114

M
Mechanical design 37, 42, 63–64, 76
Medical device 21, 28, 31–34, 37–39, 47, 73, 76, 81–86, 88, 91–92, 101–102, 104, 106, 108–109, 111–113, 115, 117–118, 152, 201, 208, 231–232, 237
Microcontroller 43–45, 47, 51, 53, 58–60, 62–63, 111
Microprocessor 40–45, 51, 53–54, 56, 59–60, 62
Motor fluctuations 2, 153–157, 160, 162–163, 166–167, 170, 172–173, 176–180, 183–184, 190–193, 195–196, 198, 200, 208, –209, 213, 215, 217–219, 228, 233, 238

Motor symptoms 2, 9, 22, 27–28, 37–39, 76, 117–119, 132, 135, 141, 156, 158, 164, 167, 170–171, 188–189, 198–199, 201, 208–209, 213, 215–217, 219, 228, 231, 233, 236–237

N
New technologies 1, 13–14, 16, 196, 200, 231, 237

P
Parkinson's disease 1–4, 12–14, 21, 24, 27, 29, 31–32, 37–38, 76, 117, 118, 152–155, 158–162, 164, 166, 172, 176, 179–180, 185, 188, 190, 194, 198, 207, 227, 231–233
Project 21–24, 27, 29, 32–34, 38–39, 62, 76, 115, 152, 216

Q
Quality management system 81, 83–84, 86, 104, 111–112, 115
Quality of life 1–2, 8–10, 12, 24, 26, 29, 154, 164, 176, 181, 187, 198–199, 213, 228, 235

R
Redesign process 37, 42, 44, 47, 64, 76
Regulatory process 81, 88, 115–116, 118
Remote monitoring 1, 14–15, 229–230, 235, 238
Report 5, 10, 12, 15, 39, 87, 90–91, 95–97, 99–102, 117–118, 123–124, 129, 131, 133, 135, 137–139, 141–143, 145, 156, 167–168, 171, 173, 175, 181, 186, 189, 192, 209–210, 212–213, 228, 233, 236
Response to treatment 1, 160, 162, 198

S
Selection of patients 14
Smartphone 22–23, 26, 28–29, 38–39, 53, 57–58, 61, 73, 113, 117–119, 122, 124–126, 129, 130

T
Therapy 2, 3, 7, 11–12, 14, 16, 175–176, 179–181, 183–184, 187, 189–190, 199, 201, 211, 213–217, 219, 228–230, 232–235, 237–238
Treatment 1–16, 27, 38, 82, 87, 98, 110, 113–114, 152–154, 156, 159–162, 166, 169–174, 176–179, 181, 185, 187–190, 192–195, 198, 200, 207–208, 212, 214, 216, 219, 229–230, 232, 236

U
User 22–23, 28, 40, 47, 51, 53–54, 61–63, 76, 88, 92, 95–96, 101–102, 104, 117–126, 130, 134, 145, 147–149, 230

W
Wearable devices 213, 218
Wearable medical device 231
Wearable sensor 21, 31, 34, 218
Wearing-off 2, 154, 156, 172–174, 176–177, 180, 182, 185, 188, 195, 209, 213, 218, 233

About the Editors

Joan Cabestany holds a Telecommunications Engineer degree (1976) from the Universitat Politècnica de Catalunya (UPC) in Barcelona, Spain. He received the Ph.D. degree in 1982 from the same university, where he is working as a Full-time Professor with the Department of Electronic Engineering.

He has been involved in research and innovation activities as a part of his career. He was responsible for the AHA ("Advanced Hardware Architectures") research group at UPC, with expertise on reconfigurable hardware, electronic system design, advanced hardware architectures, microelectronics, and VLSI design. One of the main topics of interest has been the practical application of the artificial intelligence to the functional improvement of electronic systems. More recently, he has been an active member of the ISSET ("Integrated Smart Sensors and Health Technologies") research group.

He has been a staff member and founder of the CETpD Research Center at UPC since 2005, covering activities focused on technological developments applied to the improvement and support of people with chronic diseases, such as Parkinson's disease and different conditions related with aging.

Prof. Cabestany has been responsible for several EU-funded projects. Among them, the REMPARK project on PD management and the FATE project for the accurate detection of falls in aging people are the two projects, where Prof. Cabestany has been acting as a coordinator.

He is the co-author for more than 100 research papers and communications to conferences.

Prof. Cabestany is a co-founder of the Sense4Care SL company (www.sense4care.com). The SME company is a spin-off of the UPC, commercializing relevant innovative research results. Among them, the STAT-ON™ Holter is, currently, the most relevant.

Angels Bayés Rusiñol (MD) is a Doctor in Medicine, and Neurology specialist. Her career and research activity has been focused on the study and treatment of movement disorders like Parkinson's disease (PD), Tourette syndrome, and dementias, such as Alzheimer's disease.

She has been the director of the Unit for Movement Disorders and Parkinson of TEKNON Medical Center in Barcelona (Spain) during the last 20 years.

Her main research areas are related to the implementation of holistic treatment to improve the quality of life of patients suffering from movement disorders. During the last 13 years, she has been cooperating with an engineering team at "Universitat Politècnica de Catalunya," in Barcelona, for the development and implementation of technological solutions for the support and management of PD.

She has participated in 42 research projects, with special mention of the EU-funded EDUPARK and REMPARK ones. She has authored 71 publications, including 6 books, many conferences, communications, and teaching courses.